A CHARTWELL-BRATT STU

CW01403609

Statistics and Operations Research

by I P Schagen

Department of Computer Studies
Loughborough University of Technology

Chartwell-Bratt   Studentlitteratur

©Chartwell-Bratt (Publishing & Training) Limited
ISBN 0-86238-077-4

Studentlitteratur
ISBN 91-44-23091-5

Printed in Sweden by Studentlitteratur, Lund

Contents

Acknowledgements

I would like to acknowledge the contributions of Arthur Ham, my predecessor at Loughborough, and Bob Fray, my U.S. exchange partner in 1982/3, both of whom provided some of the content and the exercises in the course. It is also important to remember all the students who have gone through the course over the years, and who have helped to give it the form it now has.

I appreciate the help and forebearance of my wife, Sandie, and children, Andrew, Paul and Claire, during the preparation of this book. Not to mention the various members of our baby-sitting group, who have provided facilities for my typewriter during many late baby-sits.

To Sandie

Introduction

Introduction

      This book developed from a course given to second-year students in the Department of Computer Studies at Loughborough University of Technology. However, a knowledge of computing is not essential for an understanding of this book, although some familiarity with programming concepts will help to appreciate the computing applications and the example programs. Those students who wish more formal instruction in programming and computing will find some useful books in the suggested further reading list at the end of Section I, on p.77.

      Throughout the book programs are introduced, and each chapter ends with some suggested computer projects. The language used is Pascal, which is becoming the standard for the serious teaching of programming. Those unfamiliar with the language should have little difficulty in following most of the examples, and may pursue the projects in any suitable language.

      Section I contains essential mathematical background for the other material in the book. Most students will probably have encountered most of this, so that for these this section is mainly important for purposes of notation and reference. It also contains a few useful programs and procedures, and is thus worth referring to even if the material is familiar.

      Section II contains some topics usually described as

"Statistics" - essentially variants on hypothesis testing. This will give a feel for statistical concepts and methods which can be expanded by further study.

Section III discusses stochastic processes, a very useful class of models for all kinds of situations where events are occurring in some random fashion. It is a particularly rich field for models of computer behaviour.

Section IV is an introduction to "deterministic" Operations Research, in which no random element is assumed to intrude. The basic ideas behind linear programming are introduced, and other problems, superficially different, are studied and found to belong to the same class of problems.

SECTION I

Mathematical Background

CHAPTER 1

Set Theory

1.1 Sets

A *set* is a group or collection of objects or symbols, of almost any type. Examples of sets are:

{ Napoleon, π, Niagara Falls }

{ 2, 4, 6, 8 }

{ A, E, I, O, U } .

Normally we place the contents of a set between curly brackets or "braces" {}. (In Pascal, to be different, we use square brackets [ ] to denote sets). The order in which objects appear is irrelevant; thus

{ 1, 2, 3 } = { 3, 1, 2 } .

Two sets are equal if all the objects in one set appear in the other and vice versa. We may define a set either by listing all its members (as in the above examples) or by giving a rule which defines what belongs to the set. For example,

{ x : x an even integer and $0 < x < 10$ } .

Read this as "The set of all x such that x is an even integer, greater than 0 and less than 10". Note that this is the same set as one of our examples above.

The objects which belong to a set are called elements or members of the set. We write:

$x \in A$ meaning "x is a member of the set A",

or $x \notin A$ meaning "x is not a member of the set A".

Two special sets are important:

The *empty set* is the set with no members and is written as $\emptyset$ or {}.

The *universal set* is the set containing everything under discussion, and we shall write it as Ω.

For example, if we were discussing cats, then we could define Ω to be the set of all cats. Other sets could be:

A = { x : $x \in \Omega$ and x has a tail } ,

B = { x : $x \in \Omega$ and x is white }

C = { x : $x \in \Omega$ and x is black with a tail } .

We can see that every member of C also belongs to A - we say C is a *subset* of A. This can be written:

$$C \subset A .$$

Alternatively, we can say A is a *superset* of C, written as

$$A \supset C .$$

1.2 Operations

We shall now consider the set *operations*, which combine sets into new sets, in the same way in which the arithmetic operations of negation, addition and multiplication combine numbers together to form new numbers.

1.2.1 <u>COMPLEMENT</u>

The complement of set A, written as $\overline{A}$ (or A' in some books), is the set containing everything not in A.

$$\overline{A} = \{ x : x \in \Omega \text{ and } x \notin A \} .$$

In the cats example above, $\overline{A}$ is the set of all cats without tails. Note that this is a so-called *unary* operation because it operates on a single set. The nearest equivalent in arithmetic is the operation of taking the negative of a number. Both operations have the property that doing them twice gets you back where you started.

$$(\overline{\overline{A}}) \quad = \quad A \quad \text{and} \quad -(-4) \quad = \quad 4.$$

1.2.2 UNION

This is a *binary* operation, operating on two sets to produce a third which contains everything in one set or the other or both. It is written as $A\cup$, and defined as

$$A\cup B \quad = \quad \{\ x : x \in A \quad \text{or} \quad x \in B\ \}\ .$$

Thus, with A and B referring to the cats example above, $A\cup B$ would be the set of all cats which are white or have tails (or both). Note that the order of the sets is irrelevant:

$$A\cup B \quad = \quad B\cup A \ .$$

Also:

$$A\cup\overline{A} \quad = \quad \Omega\ ,$$
$$A\cup A \quad = \quad A\ ,$$
$$A\cup\emptyset \quad = \quad A\ ,$$
$$A\cup\Omega \quad = \quad \Omega\ .$$

1.2.3 INTERSECTION

Another binary operation, the intersection of two sets produces a third which contains every element which is in both sets. It is written as $A\cap B$, and defined as

$$A\cap B \quad = \quad \{\ x : x \in A \text{ and } x \in B\ \}\ .$$

Thus, for the cats example, $A\cap B$ would be the set of all white cats with tails. Again:

$$A\cap B \quad = \quad B\cap A \ .$$

Also:

$$A\cap\overline{A} \quad = \quad \emptyset\ ,$$
$$A\cap A \quad = \quad A\ ,$$
$$A\cap\emptyset \quad = \quad \emptyset\ ,$$
$$A\cap\Omega \quad = \quad A\ .$$

7

1.3 Venn Diagrams

It is often an aid to understanding the relationships
between sets, especially when using the above set operations,
if a picture can be drawn. One type of picture is the Venn
diagram, which represents the universal set Ω as a rectangle
and other sets as circles within this rectangle.

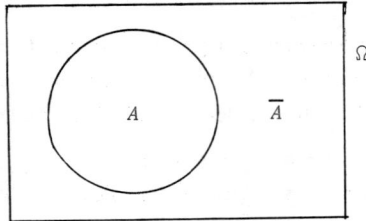

Figure 1.1

The area inside the circle represents the members of A, and
the area between the circle and the rectangle the elements
which are not in A. With more than one set, the circles
may overlap:

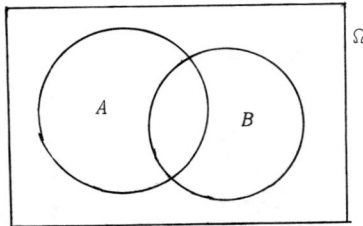

Figure 1.2

The central section is contained within both circles and
represents $A \cap B$. The whole double circle area represents
$A \cup B$. The area around the edges represents the complement
of this, written $\overline{A \cup B}$, i.e. everything not in A nor in B.
Notice that another way of writing this would be as $\overline{A} \cap \overline{B}$,
the intersection of A's complement and B's complement. If
A and B are the sets in our cats example, then $\overline{A} \cap \overline{B}$ is the
set of all cats which are not white or possessing a tail;
i.e. those which are not white and do not have a tail.

8

In situations with more than two sets we may get more complex figures - for example, with the cats problem:

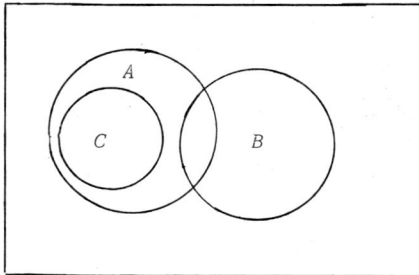

Figure 1.3

In this case C is a subset of A and does not intersect B (i.e. $C \cap B = \emptyset$). As another example, let Ω be the set of all the students at a university, and define

$$M = \{ x : x \in \Omega \text{ and } x \text{ is male } \} ,$$
$$S = \{ x : x \in \Omega \text{ and } x \text{ smokes } \} ,$$
$$C = \{ x : x \in \Omega \text{ and } x \text{ is studying computing } \} .$$

Our Venn diagram might look like:

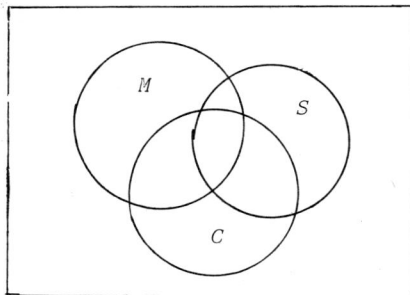

Figure 1.4

If we were asked to shade the set $(M \cap S) \cup \overline{C}$ on this diagram, the simplest way would be to translate the set operations into English. Translate $\cap$ as "and", $\cup$ as "or", and $\overline{}$ as "not". Then this set reads as: "Males and smokers or not computing students". So the set comes in two parts: males

9

and smokers, or not computing students. We shade each part,
and together they make the whole set.

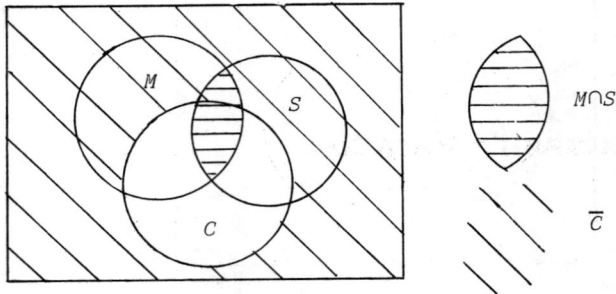

Figure 1.5

1.4 De Morgan's Law

This is an important rule in manipulating sets and
set operations. We have already seen an example earlier,
when we said that $\overline{A \cup B}$ is the same as $\overline{A} \cap \overline{B}$. De Morgan's Law
comes in two parts:

1. $\overline{A \cup B}$ = $\overline{A} \cap \overline{B}$.
2. $\overline{A \cap B}$ = $\overline{A} \cup \overline{B}$.

The second part may be translated: "The set of things not
in both A and B is the set of things which are either not
in A or not in B (or both)".

This law is quite important in programming, and
explains a very common programming bug. Sometimes people
want to write an if statement which tests, for example, if
a variable i is not equal to 1 or 2. They write:

        if (i <> 1) or (i <> 2) then ...

What's wrong with this ? If you think about it, you will
see that whatever value i has, the condition is always
true. What we want to write, using set notation, is

        if i $\notin$ {1}$\cup$\{2\}

    or   if i $\in$ $\overline{\{1\} \cup \{2\}}$

10

or   if i ∈ $\overline{\{1\}∩\{2\}}$ (de Morgan's Law)
or in Pascal:
        if (i <> 1) and (i <> 2) then ...
which is the correct condition.

1.5 Russell's Paradox

Before leaving the theory of sets, there is an interesting point which relates to the kinds of things which may belong to sets. Clearly, sets can be members of sets. We can even write
$$A = \{\emptyset\} .$$
Note that A is not equal to $\emptyset$; it is not the empty set, since it has one member (the empty set). Sets can even contain themselves. Define:
$$A = \{ x : x \text{ is a set with more than one member} \} .$$
Clearly $A \in A$.

But now we must be careful, since we run into Russell's Paradox, which warns us that we cannot allow completely arbitrary definitions of what goes into a set. Define:
$$S = \{ x : x \text{ is a set which is not a member of}$$
$$\text{itself} \} .$$
Does $S \in S$? If it does, then clearly $S \notin S$. If $S \notin S$, then equally clearly $S \in S$.

Another version of this is the Barber's Dilemna. In a certain village the barber (who is male) shaves all the males who do not shave themselves, and only these males. Who shaves the barber ?

1.6 Sets in Pascal

Set members are defined by inclusion in square brackets [ ] in Pascal. The set operations ∪ and ∩ are denoted by + and * respectively. For example:

```
program setit(input,output);
{ Program illustrating the use of sets in Pascal }
type onetoten = 1..10;
     onetotenset = set of onetoten;
var  i : onetoten;
     A,B : onetotenset;
begin
  A := [2,3,5,7]; B := [1,2,3,4,5];
  writeln; write('Input a number from 1 to 10: '); read(i);
  writeln;
  if i in A then writeln(i:2,' is in set A');
  if i in B then writeln(i:2,' is in set B');
  if i in A+B then writeln(i:2,' is in A union B')
          else writeln(i:2,' is neither in A nor in B');
  if i in A*B then writeln(i:2,' is in A intersection B');
end.
```

Sets may be constructed of any finite type (not reals), but there will be an upper limit on the number of elements in a set, which will depend on the particular computer.

1.7 Exercises

1. Let Ω be the set of all the integers from 1 to 10 inclusive. Define:

    A = { x : $x \in \Omega$ and x is prime and > 1 }
    B = { x : $x \in \Omega$ and $x \leq 5$ } ,
    C = { x : $x \in \Omega$ and x is even} .

    Write down the members of the following sets:

    a) $A \cup B \cup C$
    b) $A \cap B \cap C$
    c) $\overline{A \cup B \cup C}$
    d) $\overline{A \cap B \cap C}$

e) $(A \cup B) \cap C$.

2. A, B and C are three sets such that $C \subset A$. Draw a Venn diagram to illustrate this, and shade the following areas:
   a) $(A \cup B) \cap \overline{C}$
   b) $(C \cup B) \cap \overline{A}$
   c) $A \cap B \cap C$.

3. A, B and C are three sets such that $A \cap B = \emptyset$. Draw a Venn diagram to illustrate this, and shade the following areas:
   a) $(A \cup B) \cap \overline{C}$
   b) $(A \cup C) \cap B$
   c) $A \cap B \cap C$.

4. In general, is it true that $(A \cup B) \cap C = A \cup (B \cap C)$? Illustrate your answer using a Venn diagram.

1.8 Computer Projects

1. 9 nations are to be considered: USA, UK, France, USSR, China, Brazil, Australia, Czechoslovakia, Cuba. Three different categories are of interest: Communist nations, nations in the northern hemisphere, and island nations. Write a program, using set operations, which will enable a user to enquire about a nation and get answers to the following questions:
   a) Is it communist or in the southern hemisphere?
   b) Is it non-communist and not an island?
   c) Is it non-communist, or not an island, or in the southern hemisphere?

2. Records are to be kept on 16 students, according to whether they are male, smoke, or study computing. Write a program which will enable you to do the following:

a) Input as data the membership of each set.
b) Enquire about any student as to whether or not they belong to any set.
c) Print out the total numbers in the various combinations of sets.

(Hint: your version of Pascal may have available the function "card" which returns the number of members of a set)

CHAPTER 2

Basic Calculus

2.1 Numbers and Functions

In mathematics, various types of numbers are defined. *Integers* are the basic units of counting: 1,2,3 etc., together with the negative numbers and zero: 0,-1,-2 etc.

Rational numbers are just the ratios of two integers: e.g. 1/3, 2/5, -19/28, 14627/9436 etc. When expressed as decimals, rational numbers always have either a finite number of non-zero decimal places, or a repeating pattern of decimal digits: e.g. 0.33333..., 0.4, 3.517517517...

Irrational numbers cannot be expressed as the ratio of two integers, and have an infinite decimal representation with no regular pattern: e.g. $\sqrt{2} = 1.4142135...$, $\pi = 3.14159...$

Real numbers combine both rational and irrational, and form an infinitely dense or *uncountable* set stretching from $-\infty$ to $+\infty$. The set of all real numbers is denoted by R. An *interval* is a set of real numbers containing all the numbers between its two end-points. An open interval does not include the end-points, but a closed interval does. For example:

Open interval $\quad (2,5) = \{x:x \in R$ and $2 < x < 5\}$

Closed interval $\quad$ [2,5] $= \{x : x \in R$ and $2 \le x \le 5\}$
Half-open interval $\quad$ (2,5] $\quad \{x : x \in R$ and $2 < x \le 5\}$.
Notice how the shape of the brackets distinguishes between
open and closed intervals.

A *real function* is an operation which relates two sets
of real numbers: a *domain D* and a *range R*. Every element
of *D* is associated with a single element of *R*, although
each element of *R* may be associated with more than one
element of *D*.

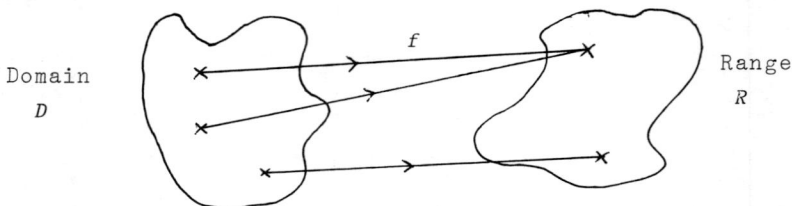

Figure 2.1

We say "*f* is a function from *D* to *R*", or write $f:D \to R$.

Example: $\quad D = $ [2,5], $R = $ [0,4], and *f* is
defined by associating with each $x \in D$
an element $y \in R$ such that
$y = (3 - x)^2$. We write $y = f(x)$.

We can graph this function, in *X* and *Y* coordinates:

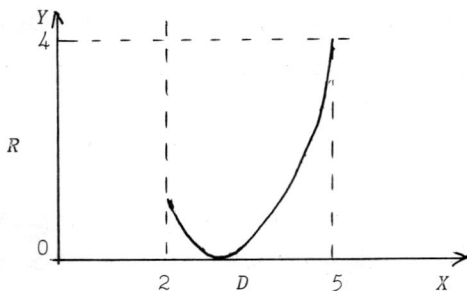

Figure 2.2

Being able to plot graphs of real functions is very
important and useful; however, it depends on the function
being "sensible", and preferably *continuous*. If $x \in D$,
then the function *f* is said to be continuous at *x* if close
to *x* the function has values close to $f(x)$. If we specify

16

a "tolerance" ε, then the function is continuous at *x* if, however small ε happens to be,we can always find a small interval around *x* within which the function values are all within ±ε of *f(x)*.

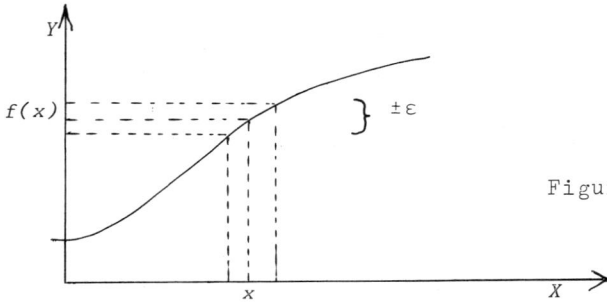

Figure 2.3

Normally, we shall deal with real functions which are continuous at all points of their domains, and can therefore be plotted as reasonably simple graphs.

A function *f* may have an *inverse function*, written as f^{-1}. It has domain *R* and range *D*, and takes the point $y \in R$ into the point $x \in D$, where $y = f(x)$. Thus $f^{-1}(f(x))$ gives *x*. For *f* to have an inverse, it must be *monotonic*, so that each $x \in D$ has a unique value of $y \in R$. Otherwise, f^{-1} would have several *x* values corresponding to the same *y* value, and thus could not be a function.

2.2 Slopes and Derivatives

When dealing with real functions, it is often necessary to discuss the slope of the function at a point or set of points. Graphically, this corresponds to the slope of the tangent line at the point.

17

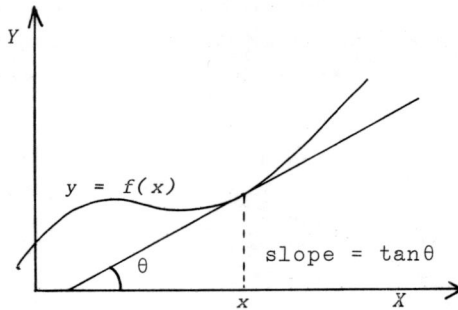

Figure 2.4

To get a value for the slope, we need to approach it indirectly, by a process of successive approximations. To get the slope of the tangent at x, approximate it by the slope of the chord-line joining the function values at x and $x+\Delta x$, where Δx is a small change in x:

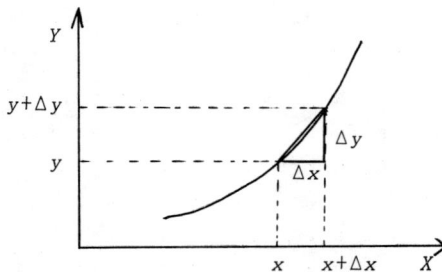

Figure 2.5

For example, if the function is defined by
$$y = f(x) = x^2,$$
then
$$y + \Delta y = (x + \Delta x)^2 = x^2 + 2x\Delta x + \Delta x^2 .$$
Subtract
$$y = x^2$$
to give
$$\Delta y = 2x\Delta x + \Delta x^2 .$$
Now, the slope of the chord is given by $\Delta y/\Delta x$. Dividing by Δx we get:
$$\frac{\Delta y}{\Delta x} = 2x + \Delta x .$$

This is an approximation to the slope of the tangent that we want. The approximation gets better as Δx gets smaller. It is easy to see that the limiting value, as $\Delta x \rightarrow 0$, is
$$\frac{\Delta y}{\Delta x} \rightarrow 2x .$$

18

This limit, for historical reasons, is called $\frac{dy}{dx}$. So we have, if $y = x^2$, the slope $\frac{dy}{dx} = 2x$ at any point x.

    This formula for the slope in fact defines another function of x, called the derivative. Another form of notation is that if the original function is $y = f(x)$, the derivative is $\frac{dy}{dx} = f'(x)$.

    The argument we gave above to get the derivative of $y = x^2$ will work for any power of x, yielding results like:

$$\text{If} \quad y = x^3 , \quad \frac{dy}{dx} = 3x^2 ;$$

$$\text{if} \quad y = x^4 , \quad \frac{dy}{dx} = 4x^3 \text{ etc.}$$

In general,   if $y = x^n$, $\frac{dy}{dx} = nx^{n-1}$.

This formula applies even to non-integer or negative values of n, for example:

$$y = \sqrt{x} \;=>\; y = x^{\frac{1}{2}} \;=>\; \frac{dy}{dx} = \tfrac{1}{2}x^{-\frac{1}{2}} = \frac{1}{2\sqrt{x}} \ .$$

$$y = \frac{1}{x^2} \;=>\; y = x^{-2} \;=>\; \frac{dy}{dx} = -2x^{-3} = \frac{-2}{x^3} \ .$$

There are some other rules for finding the derivative:

1. If $y = af(x)$, where a is a constant,
   $\frac{dy}{dx} = af'(x)$.

   Example: $y = 6x^2 \;=>\; \frac{dy}{dx} = 12x$.

2. If $y = f(x) + a$, then $\frac{dy}{dx} = f'(x)$ (The constant disappears).

3. If $y = f(ax)$, then $\frac{dy}{dx} = af'(ax)$.

   Example: $y = (2x)^3 \;=>\; \frac{dy}{dx} = 2.3(2x)^2$
   $= 24x^2$.

4. "Function of a function" rule: If $y = f(g(x))$, where f and g are both real functions, then
   $\frac{dy}{dx} = f'(g(x)).g'(x)$.

Example: $y = (2x + 3)^2$.

Let $g(x) = 2x + 3$ and $f(\ ) = (\ )^2$.

Then $g'(x) = 2$ and $f'(\ ) = 2(\ )$.

So $\dfrac{dy}{dx} = 2(2x + 3).2 = 4(2x + 3) = 8x + 12.$

2.3 Direct Applications of Differentiation

The process of obtaining the derivative of a function is known as "differentiation". The slope or derivative of a function can be considered to be the rate at which the function is changing. For example, if the Y variable relates to distance from a fixed point, and X measures time, then the slope dy/dx represents the speed or rate of change of distance with time.

Example: A ball is tossed up from the roof of a tall building, and its height above the ground is given by

$$h = 96 + 64t - 16t^2 ,$$

where t is the time in seconds. How high does it get before it starts to fall ?

The speed $v = \dfrac{dh}{dt} = 64 - 32t$ by the rules for differentiation.

The speed $v = 0$ when $t = 2$, so that the ball stops travelling upwards after 2 seconds. The height then is

$$h = 96 + 64.2 - 16.2^2 = 160 \text{ ft.}$$

2.4 Finding Maxima and Minima

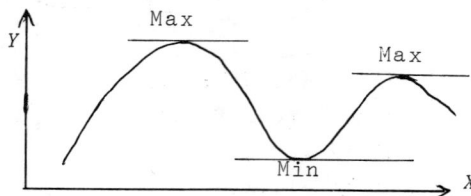

Figure 2.6

Clearly, the maximum and minimum points have slope = 0. So to find the positions of these *turning points* we differentiate, set the derivative to 0, and solve for x.

For example, if $y = 3 + 2x - 4x^2 + 2x^3$,
$$\frac{dy}{dx} = 2 - 8x + 6x^2 = 0 .$$

Solutions are $x = 1$ and $1/3$. To decide if these points are maxima or minima, we need to consider the *second derivative*, which is obtained by differentiating the function one more time, and represents the rate at which the slope is changing.
$$\frac{d^2y}{dx^2} = -8 + 12x.$$

Now if $\frac{d^2y}{dx^2} > 0$, the point is a minimum, because the slope is increasing from negative to positive,

and if $\frac{d^2y}{dx^2} < 0$, it is a maximum, because the slope is decreasing from positive to negative.

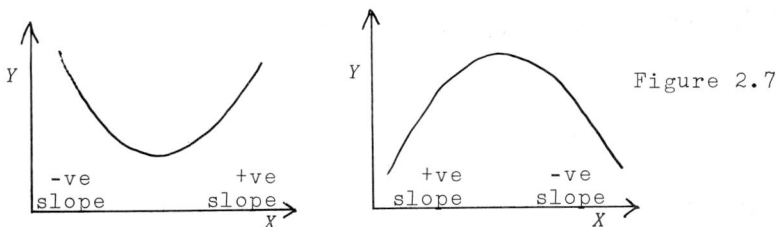

Figure 2.7

When $x = 1$, $\frac{d^2y}{dx^2} = 4$, and this is a minimum.

When $x = \frac{1}{3}$, $\frac{d^2y}{dx^2} = -4$, and this is a maximum.

Example: A can is made of a hollow cylinder of metal with circular ends, and has to contain a fixed volume. What dimensions should it have to minimise the total surface area ?

Volume $V = \pi r^2 h$ = constant.
So $h = V/\pi r^2$.
Surface Area $A = 2\pi r^2 + 2\pi rh$
$\qquad\qquad\quad = 2\pi r^2 + 2\pi r.V/\pi r^2$
$\qquad\qquad\quad = 2\pi r^2 + 2V/r$.

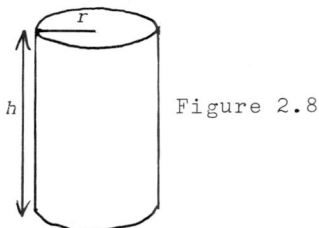

Figure 2.8

21

To find the value of r which minimises A, differentiate A as a function of r:

$$\frac{dA}{dr} = 4\pi r - 2V/r^2 = 0,$$

thus $\quad 4\pi r^3 = 2V,$

and $\quad r = (V/2\pi)^{1/3}$.

Is this a minimum ? Find the second derivative:

$$\frac{d^2A}{dr^2} = 4\pi + 4V/r^3 > 0,$$

and so it is a minimum. The value of h corresponding to the value of r which minimises A is

$$h = V/\pi r^2 = \frac{V}{\pi(V/2\pi)^{2/3}} = 2(V/2\pi)^{1/3} = 2r.$$

So the proportions of the can which minimise the amount of metal used are those which have the height of the can equal to its diameter.

2.5 Sigma Notation

The symbol $\sum$ (Greek capital sigma) is used to refer to the operation of summing several similar things. For example,

$\sum_{i=1}^{5} i \quad$ means "Sum, with i running from 1 up to 5, the values of i"

$$= 1 + 2 + 3 + 4 + 5 = 15 .$$

Also, $\sum_{i=1}^{4} i^2 = 1^2 + 2^2 + 3^2 + 4^2 = 1 + 4 + 9 + 16$
$$= 30 .$$

In general, if we have sets of n numbers like $x_1, x_2, \ldots x_n$ and $y_1, y_2, \ldots y_n$, and a constant c, the following rules for $\sum$ hold:

1. $\quad \sum_{i=1}^{n} (x_i + y_i) = \sum_{i=1}^{n} x_i + \sum_{i=1}^{n} y_i$.

2. $\quad \sum_{i=1}^{n} cx_i = c \sum_{i=1}^{n} x_i$.

3. $\quad \sum_{i=1}^{n} c = nc$.

4. $\displaystyle\sum_{i=2}^{n} (x_i - x_{i-1}) = x_n - x_1$.

2.6 Areas and Integration

Frequently, we wish to compute the area of an irregularly shaped figure. In particular, we may wish to calculate the area between the graph of a function and the x-axis, in some interval $[a,b]$.

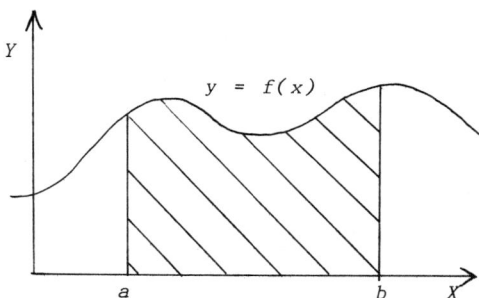

Figure 2.9

The way to approach this is to divide the interval $[a,b]$ into n small sub-intervals, each of width Δx, where $\Delta x = (b-a)/n$.

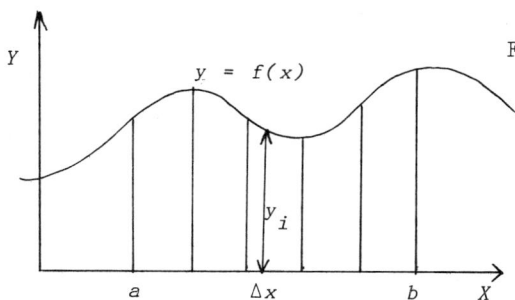

Figure 2.10

If y_i is some value in the ith sub-interval, then we can form the *Riemann sum*:

$$R_n = \sum_{i=1}^{n} y_i \Delta x .$$

As n tends to ∞, the Riemann sum R_n tends to a limit, which we shall define to be the area we want to compute, and we shall write this limit as

$$\int_a^b f(x) \, dx .$$

This is called the *definite integral* of the function f between a and b. To see how to compute values of this, we need first to discuss the *indefinite integral* of f. This is another function F, defined as

$$F(x) \quad = \quad \int_c^x f(x) \, dx, \text{ where } c \text{ is any arbitrary}$$
$$\text{fixed point.}$$

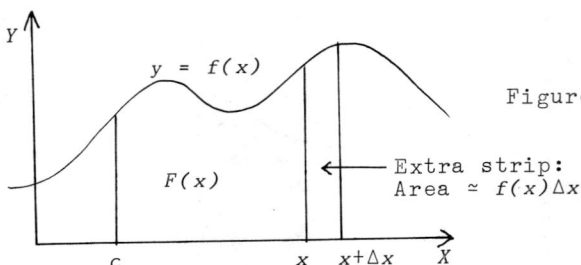

Figure 2.11

If we increase x to $x+\Delta x$, it is fairly clear that
$$F(x+\Delta x) \quad \simeq \quad F(x) + f(x)\Delta x .$$
So the derivative of F is

$$F'(x) \quad = \quad \lim_{\Delta x \to 0} \frac{F(x+\Delta x) - F(x)}{\Delta x} \quad = \quad f(x) .$$

The *Fundamental Theorem of Calculus* states that:

1. If $F(x) = \int_c^x f(x) \, dx$,

    then its derivative $F'(x) = f(x)$.

2. $\int_a^b f(x) \, dx = F(b) - F(a)$.

Thus integration is the inverse operation to differentiation; if we can "anti-differentiate" a function, then we can compute areas etc. We normally write the indefinite integral as:

$$\int f(x) \, dx,$$

which is the function which, when differentiated, gives $f(x)$.

24

Example: $\int (2x + 3x^2)\, dx = ?$

A bit of thought shows that differentiating the function
$y = x^2 + x^3$ gives the above function, so

$$\int (2x + 3x^2)\, dx = x^2 + x^3 .$$

In fact, differentiating any function like
$y = x^2 + x^3 + c$ (c is any constant) gives the same result,
so we usually write for the indefinite integral:

$$\int (2x + 3x^2)\, dx = x^2 + x^3 + c .$$

To find definite integrals, e.g. $\int_{2}^{4} (2x + 3x^2)\, dx$, we

just substitute in the indefinite integral. We usually
write:

$$\int_{2}^{4} (2x + 3x^2)\, dx = \left[x^2 + x^3 \right]_{2}^{4} .$$

This means "indefinite integral evaluated at 4, minus the
value evaluated at 2".

$$\left[x^2 + x^3 \right]_{2}^{4} = (4^2 + 4^3) - (2^2 + 2^3)$$
$$= 80 - 12 = \underline{68} .$$

2.7 Rules for Integration

1. $\int (f(x) + g(x))\, dx = \int f(x)\, dx + \int g(x)\, dx .$

2. $\int cf(x)\, dx = c\int f(x)\, dx .$

3. $\int_{a}^{b} f(x)\, dx = -\int_{b}^{a} f(x)\, dx .$

   (Reversing the direction of integration changes
   the sign of the integral).

4. $\int_{a}^{b} f(x)\, dx = \int_{a}^{c} f(x)\, dx + \int_{c}^{b} f(x)\, dx .$

5.   The general rule for integrating powers of x is the reverse of the rule for differentiating them:

$$\int x^n \, dx = \frac{x^{n+1}}{n+1} + C \, .$$

This is true for all values of n, except $n = -1$. We shall consider this case presently, in section 2.9, when we deal with the natural logarithmic function.

2.8 Applications of Integration

Example: Find the area between the curve $y = 8x - 2x^2$ and the x-axis.

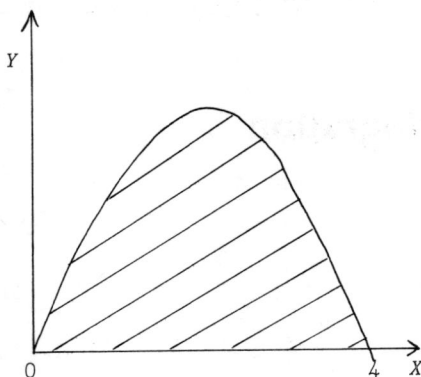

Figure 2.12

The curve crosses the x-axis at $x = 0$ and $x = 4$.

Required area $= \int_0^4 (8x - 2x^2) \, dx = \left[4x - \frac{2x^3}{3} \right]_0^4$

26

$$= 64 - \frac{128}{3} = \frac{64}{3} \quad .$$

Example: A rocket travels so that its acceleration after
t seconds is $20t$ ft/sec^2. How far has it gone
after 3 seconds, if it starts from rest ?

Acceleration $a = \frac{dv}{dt} = 20t$.

So $v = \int a \, dt = \int 20t \, dt = 10t^2 + C$.

Now when $t = 0$, $v = 0$, so C must be equal to 0.
Therefore $v = \frac{ds}{dt} = 10t^2$.

So $s = \int v \, dt = \int 10t^2 \, dt = \frac{10t^3}{3} + C$.

Now when $t = 0$, $s = 0$, so C must be equal to 0.
Therefore $s = \frac{10t^3}{3}$.

So when $t = 3$, $s = 90$ ft.

2.9 Natural Logarithmic and Exponential Functions

We showed earlier that the formula for integrating
a power of x does not work for x^{-1}; in other words, we do
not know how to integrate $y = 1/x$.

Let us define a new function:

$$\ell n(x) = \int_1^x \frac{1}{x} \, dx \quad .$$

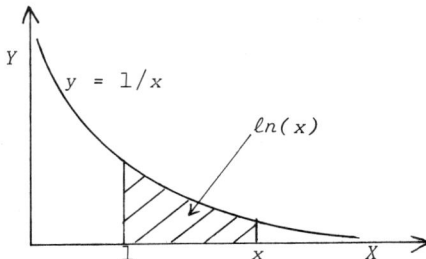

Figure 2.13

Clearly, $\ell n(x)$ is an indefinite integral of $1/x$. What is more interesting is to examine the properties of $\ell n(x)$.

Consider $y = \ell n(cx)$, and differentiate this:

$$\frac{dy}{dx} = \frac{1}{cx} . c \quad \text{by the "function of a function" rule}$$

$$= \frac{1}{x} = \text{the derivative of } \ell n(x).$$

So the two functions, $\ell n(x)$ and $\ell n(cx)$, have the same slope for any value of x, and must therefore only differ by a constant.

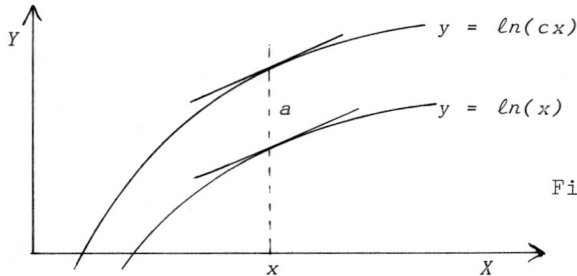

Figure 2.14

So $\ell n(cx) = \ell n(x) + a$.
When $x = 1$, $\ell n(c) = \ell n(1) + a$,
so that $a = \ell n(c)$.
Therefore $\ell n(cx) = \ell n(x) + \ell n(c)$.

This means that the function ℓn has the property of a logarithm, of converting multiplication into addition. To find the base of the logarithm, we need to find a value of x such that $\ell n(x) = 1$;

$$\text{i.e.} \quad \int_1^x \frac{1}{x} \, dx = 1 .$$

This value is about $2.718...$, and is given the name e, the base of *natural logarithms*, and we sometimes write $\log_e x$ instead of $\ell n(x)$.

Example: $\int_0^2 \frac{dx}{1+x} = \left[\ell n(1+x) \right]_0^2 = \ell n(3) - \ell n(1)$

$$= \underline{1.099} .$$

28

The inverse function of *ln* is called *exp*, or the *natural exponential* function. It is defined so that if

$$y = ln(x) , \quad x = exp(y) .$$

Graphically, the two functions look like:

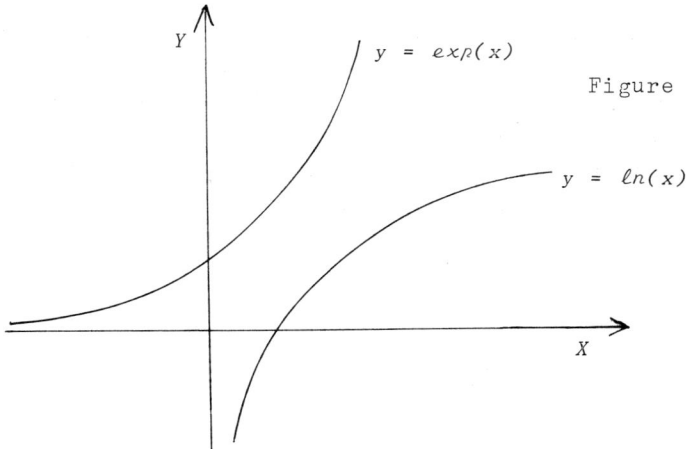

$y = exp(x)$

Figure 2.15

$y = ln(x)$

$y = exp(x)$ is also written $y = e^x$, as it is equivalent to raising e to the power of x (as in taking anti-logarithms).

The function $y = e^x$ has the property that it is its own derivative:

$$\frac{dy}{dx} = e^x .$$

So the reverse holds: $\int e^x \, dx = e^x .$

This function is useful in modelling growth or decay. For example, if a quantity y is growing so that its rate of growth is proportional to y itself at time t:

$$\frac{dy}{dt} = ky ,$$

then a functional form of y which satisfies this is

$$y = y_0 e^{kt} , \text{ where } y_0 \text{ is the value at}$$
$$t = 0 .$$

2.10 Numerical Integration

Not all functions have indefinite integrals which may be written down exactly - for example, there is no exact indefinite integral for the function $y = exp(-x^2/2)$, which we shall meet in chapter 4. To compute a definite integral, we shall need to approximate the required area. Two simple methods are available to approximate

$$\int_a^b f(x) \, dx \quad .$$

2.10.1 THE TRAPEZOIDAL RULE

Divide the interval $[a,b]$ into n equal intervals, and evaluate the function at the end-points of these intervals, giving $n+1$ values $y_0, y_1, \ldots y_n$.

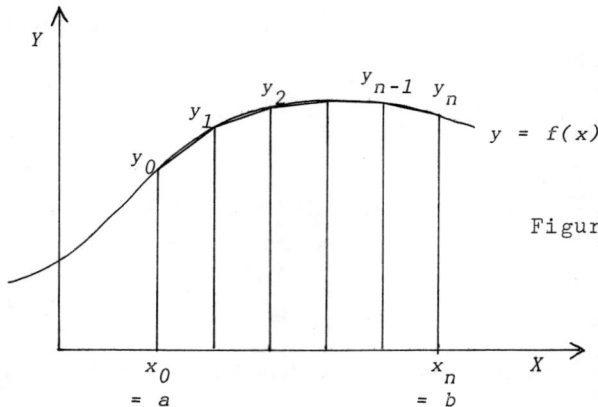

Figure 2.16

Let Δx = width of each strip = $(b-a)/n$. Approximate the area under the function in each strip by the area of the corresponding trapezium; e.g. in the sub-interval $[x_0, x_1]$, the area = $\frac{1}{2}(y_0 + y_1)\Delta x$.

Therefore the total area

$$= \tfrac{1}{2}(y_0 + y_1)\Delta x + \tfrac{1}{2}(y_1 + y_2)\Delta x + \ldots + \tfrac{1}{2}(y_{n-1} + y_n)\Delta x$$

$$= (\tfrac{1}{2}y_0 + y_1 + y_2 + \ldots + y_{n-1} + \tfrac{1}{2}y_n)\Delta x$$

$$\simeq \int_a^b f(x)\ dx\ .$$

For example, if we wish to estimate $\int_0^1 exp(-x^2/2)\ dx$ with a division into 4 intervals. $\Delta x = 0.25$.

$$
\begin{array}{llll}
x_0 & = & 0.0 & \qquad y_0 & = & 1.0 \\
x_1 & = & 0.25 & \qquad y_1 & = & 0.9692 \\
x_2 & = & 0.5 & \qquad y_2 & = & 0.8825 \\
x_3 & = & 0.75 & \qquad y_3 & = & 0.7548 \\
x_4 & = & 1.0 & \qquad y_4 & = & 0.6065 \\
\end{array}
$$

By the trapezoidal rule, the required integral $\simeq 0.8524$. (From tables, the true value is 0.8555).

Here is a listing of a simple program to integrate a function numerically using the trapezoidal rule:

```
program trapezoidal(input,output);
{ Program for numerical integration by trapezoidal rule }
const maxvals = 50;
type  ordinates = array[ 0..maxvals ] of real;
      posint = 0..maxint;
var   i,nsub : posint;
      lower,upper,integral,x,dx : real;
      y : ordinates:

function trapint(var y : ordinates; dx : real;
                 n : posint) : real;
  var intval : real;
      i : posint;
  begin
    intval := 0.5*(y[ 0 ]+ y[ n ]);
    for i := 1 to n-1 do
      intval := intval + y[ i ];
    trapint := intval*dx;
  end;

begin
  writeln; write('Input lower limit of integration: ');
  read(lower); writeln;
  write('Input upper limit of integration: ');
  read(upper); writeln;
  write('Input no. of sub-intervals: ');
  read(nsub); writeln;
  x := lower; dx := (upper-lower)/nsub;
```

```
for i := 0 to nsub do
begin
  write('Input y value at x = ',x:8:4,' - ');
  read(y[ i ]); writeln;
  x := x + dx;
end;
integral := trapint(y,dx,nsub);
writeln; writeln('Value of integral = ',integral:10:4);
end.
```

2.10.2 SIMPSON'S RULE

This method assumes n is even, and the strips are
all of the same width. We take them in pairs, and fit
a quadratic approximating function to the three adjacent
y values.

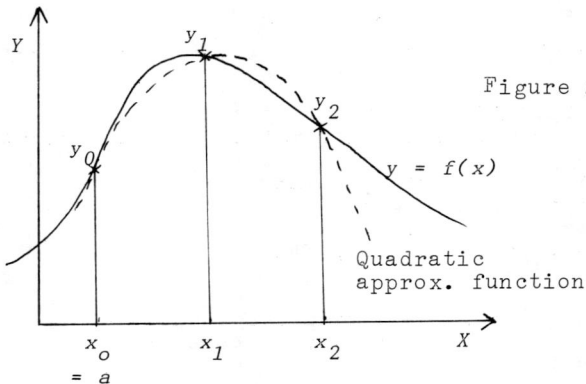

Figure 2.17

We thus end up with $n/2$ quadratic approximating functions
to cover the whole interval of integration.

The area under the approximating function for the
first pair of strips is

$$\frac{\Delta x}{3} (y_0 + 4y_1 + y_2) \quad .$$

So the total area, taking strips in pairs, is

$$\frac{\Delta x}{3} (y_0 + 4y_1 + y_2) + \frac{\Delta x}{3} (y_2 + 4y_3 + y_4) + \cdots$$

$$= \frac{\Delta x}{3} (y_0 + 4y_1 + 2y_2 + 4y_3 + 2y_4 + \cdots + 4y_{n-1} + y_n)$$

$$\simeq \int_a^b f(x) \, dx \quad .$$

32

For our example, with $n = 4$, the approximate integral
using Simpson's Rule is 0.8556.

2.11 Exercises

1. Use the method of Δx and Δy (as on p.16) to derive the
    derivatives of the following functions:
    a) $y = 2x^4$ b) $y = 1/x$ c) $y = \sqrt{x}$.

2. Use the rules of differentiation to obtain the
    derivatives of the following functions:
    a) $y = \sqrt{9x + 2x^2}$ b) $y = 1/(3x + 1)$
    c) $y = x^3 + 2x^2 - 1$ d) $y = \ln(1/x)$.

3. A jogger runs so that their distance from home, x
    miles, after t hours is given by $x = 15t - t^2$.
    How far from home are they when they turn round ?
    How long is it before they return home ?

4. A farmer buys 1000 metres of fencing. Prove by
    calculus that the maximum rectangular area that he
    can totally enclose with it forms a square.

5.

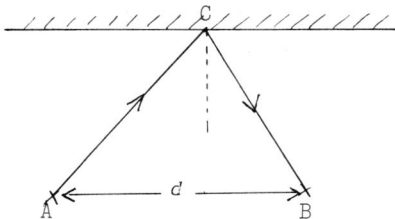

Figure 2.18

Light always travels by the shortest path. In Figure
2.18 it is reflected from A to B via a mirror. If A
and B are a distance d apart, show by calculus that
the point of reflection C is such that the angle of
reflection equals the angle of incidence.

33

6. A company has a total of £10,000 to invest. If it invests £x thousand in project A, the profit is £10x thousand. Investing £y thousand in project B gives a profit of £50$\sqrt{y}$ thousand. How should the company invest its money to maximise the total profit ?

7. Find the following indefinite integrals:

a) $\int (2x + 8x^5)\ dx$

b) $\int \dfrac{dx}{2x^2}$

c) $\int (1 + x)^{-1}\ dx$

d) $\int exp(-2x)\ dx$.

8. Evaluate the following definite integrals:

a) $\int_0^2 (8 - 3x^2)\ dx$

b) $\int_{-1}^1 6x\ dx$

c) $\int_0^4 x^2\ dx$

d) $\int_0^1 exp(1+x)\ dx$.

9. Find the area between the curve $y = 10x - x^2$ and the x-axis.

10. Acceleration under gravity is 32 ft/sec^2. A ball is tossed over a 100 ft cliff with an upward vertical velocity of 20 ft/sec. How long will it be before it reaches the bottom ?

11. Use both the Trapezoidal Rule and Simpson's Rule with 6 intervals to compute approximations to

$$\int_1^2 ln(x^2)\ dx\ .$$

2.12 Computer projects

1. Write a computer program which will accept as input the coefficients of a polynomial of x, and output the coefficients of its derivative. The program should

also be able to compute the values of the function
and its derivative at any specified set of points.

2. Write a computer program (or extend the above) to
    output the coefficients of the indefinite integral of
    a polynomial. The program should also be able to
    compute the value of the definite integral between any
    two specified points.

3. Write a computer program to approximate the definite
    integral of a function using Simpson's Rule.

CHAPTER 3

Matrices and Vectors

3.1 Definitions

A *matrix* is a rectangular table or array of numbers, such as

$$A = \begin{pmatrix} 2 & 4 & -1 & 8 \\ 3 & 9 & -6 & 2 \\ -1 & 4 & 0 & 5 \end{pmatrix}$$

The above matrix has 3 rows and 4 columns, and is said to be a 3×4 matrix.

A *vector* is a column of numbers, such as

$$\underset{\sim}{b} = \begin{pmatrix} 3 \\ 2 \\ -1 \\ 4 \end{pmatrix}$$

This is a 4×1 matrix, or 4 element column vector. We shall adopt the convention of using upper-case letters to refer to matrices, and lower-case letters with the symbol ~ underneath to refer to vectors. We shall use lower-case letters by themselves to refer to ordinary numbers, or *scalars*.

There are various operations that can be performed on matrices and vectors.

3.1.1 SCALAR MULTIPLICATION

This operation just involves multiplying each element of a matrix or vector by the same scalar.

$$e.g. \quad 2A \quad = \quad 2\begin{pmatrix} 2 & 4 & -1 & 8 \\ 3 & 9 & -6 & 2 \\ -1 & 4 & 0 & 5 \end{pmatrix} \quad = \quad \begin{pmatrix} 4 & 8 & -2 & 16 \\ 6 & 18 & -12 & 4 \\ -2 & 8 & 0 & 10 \end{pmatrix}$$

$$3\underset{\sim}{b} \quad = \quad 3\begin{pmatrix} 3 \\ 2 \\ -1 \\ 4 \end{pmatrix} \quad = \quad \begin{pmatrix} 9 \\ 6 \\ -3 \\ 12 \end{pmatrix}$$

3.1.2 MATRIX ADDITION AND SUBTRACTION

To add or subtract two matrices, it is necessary that they have exactly the same number of rows and columns. Then it is just a matter of adding or subtracting the corresponding elements of the two matrices.

$$e.g. \quad A \quad = \quad \begin{pmatrix} 2 & 4 & -1 & 8 \\ 3 & 9 & -6 & 2 \\ -1 & 4 & 0 & 5 \end{pmatrix} \quad B \quad = \quad \begin{pmatrix} 1 & 5 & 2 & -3 \\ 0 & 1 & 4 & -3 \\ 2 & 1 & 6 & 1 \end{pmatrix}$$

$$A+B \quad = \quad \begin{pmatrix} 3 & 9 & 1 & 5 \\ 3 & 10 & -2 & -1 \\ 1 & 5 & 6 & 6 \end{pmatrix} \quad A-B \quad = \quad \begin{pmatrix} 1 & -1 & -3 & 11 \\ 3 & 8 & -10 & 5 \\ -3 & 3 & -6 & 4 \end{pmatrix}$$

3.1.3 TRANSPOSITION

The *transpose* of a matrix is defined to be the same matrix with rows and columns interchanged. The transpose of A is written as A' (in some books as A^T).

e.g. A = $\begin{pmatrix} 2 & 4 & -1 & 8 \\ 3 & 9 & -6 & 2 \\ -1 & 4 & 0 & 5 \end{pmatrix}$ A' = $\begin{pmatrix} 2 & 3 & -1 \\ 4 & 9 & 4 \\ -1 & -6 & 0 \\ 8 & 2 & 5 \end{pmatrix}$

The transpose of an $m \times n$ matrix is therefore an $n \times m$ matrix. The transpose of an $m \times 1$ vector $\underset{\sim}{b}$ will be a $1 \times m$ *row vector* $\underset{\sim}{b'}$

e.g. $\underset{\sim}{b}$ = $\begin{pmatrix} 3 \\ 2 \\ -1 \\ 4 \end{pmatrix}$ $\underset{\sim}{b'}$ = $(\ 3 \quad 2 \quad -1 \quad 4\ )$

3.1.4 SCALAR PRODUCT

The *scalar product* of a row vector and a column vector is obtained by multiplying corresponding elements of each vector and summing to give a scalar.

e.g. $\underset{\sim}{a}$ = $\begin{pmatrix} 1 \\ 3 \\ 6 \\ 0 \end{pmatrix}$ $\underset{\sim}{b}$ = $\begin{pmatrix} 3 \\ 2 \\ -1 \\ 4 \end{pmatrix}$

$\underset{\sim}{a'} \cdot \underset{\sim}{b}$ = $(\ 1 \quad 3 \quad 6 \quad 0\ )$ $\begin{pmatrix} 3 \\ 2 \\ -1 \\ 4 \end{pmatrix}$

$= 1 \times 3 + 3 \times 2 + 6 \times (-1) + 0 \times 4 = \underline{3}$.

3.1.5 MATRIX MULTIPLICATION

The rule for multiplying two matrices together assumes that the number of columns of the first matrix is equal to the number of rows of the second matrix. Then a new matrix is formed whose (i,j)th element is given by the scalar product of the ith row of the first matrix with the jth column of the second matrix.

For example, if $C = AB$, where

$$A = \begin{pmatrix} 2 & 4 & -1 & 8 \\ 3 & 9 & -6 & 2 \\ -1 & 4 & 0 & 5 \end{pmatrix} \qquad B = \begin{pmatrix} 2 & 0 \\ 1 & 1 \\ 3 & 1 \\ 0 & 2 \end{pmatrix}$$

then the value in row 2, column 1, of C is given by the scalar product of A's row 2 with B's column 1:

$$c_{21} = \begin{pmatrix} 3 & 9 & -6 & 2 \end{pmatrix} \begin{pmatrix} 2 \\ 1 \\ 3 \\ 0 \end{pmatrix} = -3 \;.$$

Applying this rule to get all the elements of C, we find:

$$C = \begin{pmatrix} 2 & 4 & -1 & 8 \\ 3 & 9 & -6 & 2 \\ -1 & 4 & 0 & 5 \end{pmatrix} \begin{pmatrix} 2 & 0 \\ 1 & 1 \\ 3 & 1 \\ 0 & 2 \end{pmatrix} = \begin{pmatrix} 5 & 19 \\ -3 & 7 \\ 6 & 14 \end{pmatrix}$$

In this case we have multiplied a 3×4 matrix by a 4×2 matrix, giving a 3×2 matrix as a result. In general, multiplying an $m{\times}n$ matrix by an $n{\times}k$ matrix gives an $m{\times}k$ matrix as a result.

We can define the operation of matrix multiplication algebraically as follows:

Let a_{ij} be the value of the (i,j)th element of A,
$b_{j\ell}$ be the value of the (j,ℓ)th element of B

39

and $c_{i\ell}$ be the value of the (i,ℓ)th element of C.

Then if $C = AB$,

$$c_{i\ell} = \sum_{j=1}^{n} a_{ij} b_{j\ell} \ .$$

We can use this definition to produce a Pascal procedure for multiplying together two matrices:

```
const maxrowcol = 10;
type matrix = array[1..maxrowcol,1..maxrowcol] of real;
     posint = 0..maxint;

procedure matmult(var a,b,c : matrix; m,n,k : posint);
var i,j,ℓ : posint;
begin
  for i := 1 to m do
    for ℓ := 1 to k do
    begin
      c[i,ℓ] := 0.0;
      for j := 1 to n do
        c[i,ℓ] := c[i,ℓ] + a[i,j]*b[j,ℓ];
    end;
end;
```

An important point to notice about matrix multiplication is that, unlike ordinary multiplication, the order is important.

$AB \neq BA$ (in fact, BA may be impossible).

For example, if $A = \begin{pmatrix} 2 & 1 & 3 \\ 1 & 0 & 2 \end{pmatrix}$ $B = \begin{pmatrix} 1 & -1 \\ 1 & 0 \\ 2 & -1 \end{pmatrix}$

then $AB = \begin{pmatrix} 2 & 1 & 3 \\ 1 & 0 & 2 \end{pmatrix} \begin{pmatrix} 1 & -1 \\ 1 & 0 \\ 2 & -1 \end{pmatrix} = \begin{pmatrix} 9 & -5 \\ 6 & -3 \end{pmatrix}$

while $BA = \begin{pmatrix} 1 & -1 \\ 1 & 0 \\ 2 & -1 \end{pmatrix} \begin{pmatrix} 2 & 1 & 3 \\ 1 & 0 & 2 \end{pmatrix} = \begin{pmatrix} 1 & 1 & 1 \\ 2 & 1 & 3 \\ 3 & 2 & 4 \end{pmatrix}$

The sizes of the resulting matrices are different, as well as their values.

Note also what happens when you transpose a product:

$$(AB)' = B'A' \ .$$

3.2 Matrix Inversion

A matrix is *square* if it has as many rows as columns. An $m \times m$ square matrix is an *identity matrix* if it has values of 1 for all its main diagonal elements and 0 elsewhere. For example,

$$I \; = \; \begin{pmatrix} 1 & 0 & 0 \\ 0 & 1 & 0 \\ 0 & 0 & 1 \end{pmatrix} \quad \text{is a 3×3 identity matrix,}$$

or *unit matrix*.

The symbol I is normally used for an identity matrix, and such a matrix has the property that multiplication by it (in front or behind) leaves any other matrix unchanged.

e.g.
$$\begin{pmatrix} 1 & 0 & 0 \\ 0 & 1 & 0 \\ 0 & 0 & 1 \end{pmatrix} \begin{pmatrix} 1 & -1 \\ 1 & 0 \\ 2 & -1 \end{pmatrix} \; = \; \begin{pmatrix} 1 & -1 \\ 1 & 0 \\ 2 & -1 \end{pmatrix}$$

An $m \times m$ square matrix A is said to possess an *inverse* if there exists another matrix A^{-1} such that

$$AA^{-1} \; = \; I \; = \; A^{-1}A \; .$$

For example, does the matrix $A \; = \; \begin{pmatrix} 1 & 2 \\ 3 & 4 \end{pmatrix}$ have an inverse ?

We can produce the matrix $A^{-1} \; = \; \begin{pmatrix} -2 & 1 \\ 3/2 & -\frac{1}{2} \end{pmatrix}$

and show $AA^{-1} \; = \; \begin{pmatrix} 1 & 2 \\ 3 & 4 \end{pmatrix} \begin{pmatrix} -2 & 1 \\ 3/2 & -\frac{1}{2} \end{pmatrix} \; = \; \begin{pmatrix} 1 & 0 \\ 0 & 1 \end{pmatrix}$

and $A^{-1}A \; = \; \begin{pmatrix} -2 & 1 \\ 3/2 & -\frac{1}{2} \end{pmatrix} \begin{pmatrix} 1 & 2 \\ 3 & 4 \end{pmatrix} \; = \; \begin{pmatrix} 1 & 0 \\ 0 & 1 \end{pmatrix}$

If a matrix A has an inverse, then that inverse is unique. Suppose B were another inverse of A, not equal to A^{-1}, then

$$AB \; = \; I \; .$$

Multiply both sides of this by A^{-1}, in front:

$$A^{-1}AB \; = \; A^{-1}I$$

$$\Rightarrow \quad I B \ = \ A^{-1}$$
$$\Rightarrow \quad B \ = \ A^{-1} \ . \quad \text{So } A^{-1} \text{ is unique.}$$

Any square matrix which does not possess an inverse is said to be *singular*.

3.3 Computing the Inverse of a Matrix

The next problem is to compute the inverse of a non-singular square matrix. Consider first the associated problem of finding a solution to the simultaneous equations:
$$x \ + \ 2y \ = \ 3$$
$$3x \ + \ 4y \ = \ 5 \ .$$
We could write this in matrix form:
$$A \underset{\sim}{x} \ = \ \underset{\sim}{b} \ ,$$
where $\quad A \ = \ \begin{pmatrix} 1 & 2 \\ 3 & 4 \end{pmatrix} , \quad \underset{\sim}{x} \ = \ \begin{pmatrix} x \\ y \end{pmatrix} \quad$ and $\quad \underset{\sim}{b} \ = \ \begin{pmatrix} 3 \\ 5 \end{pmatrix} .$

Multiplying both sides of the matrix equation by A^{-1} we get:
$$A^{-1} A \underset{\sim}{x} \ = \ \underset{\sim}{x} \ = \ A^{-1} \underset{\sim}{b} \ .$$
$$\text{So} \quad \underset{\sim}{x} \ = \ \begin{pmatrix} -2 & 1 \\ 3/2 & -\frac{1}{2} \end{pmatrix} \begin{pmatrix} 3 \\ 5 \end{pmatrix} \ = \ \begin{pmatrix} -1 \\ 5 \end{pmatrix} \ .$$

The solution is $x = -1$, $y = 5$. So, in a sense, solving simultaneous equations and inverting matrices are the same problem.

Suppose we just wrote $\underset{\sim}{b} \ = \ \begin{pmatrix} b_1 \\ b_2 \end{pmatrix}$, so the equations became:
$$x \ + \ 2y \ = \ b_1 \qquad (1)$$
$$3x \ + \ 4y \ = \ b_2 \qquad (2)$$
Rewrite equation (1) as:
$$x \ = \ b_1 \ - \ 2y$$
and substitute for x in equation (2):
$$3b_1 \ - \ 6y \ + \ 4y \ = \ b_2$$
$$\Rightarrow \quad 3b_1 \ - \ 2y \ = \ b_2 \ .$$

We have replaced b_1 on the right-hand side by x, and in matrix form the equations now look like:

$$\begin{pmatrix} 1 & -2 \\ 3 & -2 \end{pmatrix} \begin{pmatrix} b_1 \\ y \end{pmatrix} = \begin{pmatrix} x \\ b_2 \end{pmatrix} .$$

We say that we have *pivoted* on the first diagonal element of A, to produce a new matrix A^*. Pivoting on the second diagonal element (replacing b_2 on the right-hand side by y) will give us A^{-1}.

In the more general case, with m equations in m variables $x_1, x_2, \ldots x_m$, we have:

$$
\begin{aligned}
a_{11}x_1 + a_{12}x_2 + \cdots + a_{1m}x_m &= b_1 \\
a_{21}x_1 + a_{22}x_2 + \cdots + a_{2m}x_m &= b_2 \\
&\vdots \\
a_{m1}x_1 + a_{m2}x_2 + \quad + a_{mm}x_m &= b_m
\end{aligned}
$$

Pivoting on the ith diagonal element of this matrix, replacing b_i by x_i on the right-hand side, we get, for the ith equation:

$$x_i = \frac{b_i}{a_{ii}} - \frac{a_{i1}x_1}{a_{ii}} - \cdots - \frac{a_{im}x_m}{a_{ii}} .$$

For other equations, say the jth, we replace x_i by the above equation to get:

$$a_{j1}x_1 + \cdots + a_{jm}x_m - \frac{a_{ji}a_{i1}x_1}{a_{ii}} - \cdots - \frac{a_{ji}a_{im}x_m}{a_{ii}}$$
$$+ \frac{a_{ji}b_i}{a_{ii}} = b_j .$$

So the elements of the new pivoted matrix A^* are given by:

For row i:
$$a^*_{ii} = 1/a_{ii}$$
$$a^*_{ik} = -a_{ik}/a_{ii} \quad \text{for } k \neq i .$$

For row $j \neq i$:
$$a^*_{ji} = a_{ji}/a_{ii}$$
$$a^*_{jk} = a_{jk} - a_{ji}a_{ik}/a_{ii} \quad \text{for } k \neq i .$$

If we repeat this pivot operation m times, once for each diagonal element of A, we will finally transform A into

A^{-1}, because we will have replaced all the $\underset{\sim}{b}$ values on the right-hand side by $\underset{\sim}{x}$ values, and can write

$$A^{-1}\underset{\sim}{b} = \underset{\sim}{x} \quad \text{instead of} \quad A\underset{\sim}{x} = \underset{\sim}{b} \ .$$

The following Pascal program is designed to invert a square matrix using the above pivoting method:

```
program invert(input,output);
{  Program to read in and invert a square matrix }
const maxrowcol = 10;
type matrix = array[1..maxrowcol,1..maxrowcol] of real;
     posint = 0..maxint;
var  a,ainverse : matrix;
     m : posint;
     singular : boolean;

procedure matread(var a : matrix; var m : posint);
{ Procedure to read in values of a square matrix }
  var i,j : posint;
  begin
    writeln; write('Input size of matrix: ');
    read(m); writeln;
    for i := 1 to m do
    begin
      writeln('Input elements of row',i:3);
      for j := 1 to m do read(a[i,j]);
    end;
  end;

procedure invert(var a,ainverse : matrix; m : posint;
                 var singular : boolean);
{ Procedure to invert a square matrix by pivoting }
  var astar : matrix;
      i,j,k : posint;
  begin
    for i := 1 to m do
      for j := 1 to m do ainverse[i,j] := a[i,j];
    i := 1; singular := false;
    while (i <= m) and not singular do
    begin
      if abs(ainverse[i,i]) > 1.0e-6 then
      begin
        for j := 1 to m do
        begin
          if j = i then
            for k := 1 to m do
              if k = i
              then astar[i,i] := 1.0/ainverse[i,i]
              else
                astar[i,k] := -ainverse[i,k]/
                                   ainverse[i,i]
```

```
                else
                   for k := 1 to m do
                      if k = i
                         then astar[j,i] := ainverse[j,i]/
                                            ainverse[i,i]
                         else
                            astar[j,k] := ainverse[j,k]
                             - ainverse[j,i]*ainverse[i,k]/
                                            ainverse[i,i];
            end;
            for j := 1 to m do
               for k := 1 to m do ainverse[j,k] := astar[j,k];
            i := i + 1;
         end
         else begin
            singular := true;
            writeln('Singular matrix');
         end;
      end;
   end;

procedure matprint(var p : matrix; m : posint);
{ Matrix printing routine }
   var i,j : posint;
   begin
      writeln;
      for i := 1 to m do
      begin
         for j := 1 to m do write(p[i,j]:7:2);
         writeln;
      end;
   end;

{ Main routine }
begin
   matread(a,m);
   invert(a,ainverse,m,singular);
   writeln; writeln('Inverse matrix - ');
   matprint(ainverse,m);
end.
```

Note that we detect a singular matrix by the fact that we are trying to divide by $a_{ii} = 0$ at a certain point.

For example, suppose we were trying to solve:

$$x + 2y = 3 \qquad A = \begin{pmatrix} 1 & 2 \\ 2 & 4 \end{pmatrix}$$
$$2x + 4y = 5$$

Pivoting on the first diagonal element of A gives

$$A^* = \begin{pmatrix} 1 & -2 \\ 2 & 0 \end{pmatrix}$$

Trying to pivot on the second diagonal element leads to a division by zero, so the matrix is singular - this pair of equations cannot be solved.

3.4 Exercises

1. Given that $A = \begin{pmatrix} 2 & 1 & 0 & 4 \\ 3 & -1 & 2 & 1 \end{pmatrix}$, $\underset{\sim}{a} = \begin{pmatrix} 1 \\ 2 \\ 3 \end{pmatrix}$

   $B = \begin{pmatrix} -2 & 1 \\ -1 & 2 \\ 3 & 3 \end{pmatrix}$ and $\underset{\sim}{b} = \begin{pmatrix} 1 \\ 1 \end{pmatrix}$

   find the following (or show that they do not exist):
   a) $A' + B$ b) $\underset{\sim}{b}'A$ c) BA
   d) $B'\underset{\sim}{a}$ e) AB

2. Compute the following matrix and vector products:
   a) $(1 \quad 2 \quad 3) \begin{pmatrix} 4 \\ 5 \\ 6 \end{pmatrix}$

   b) $\begin{pmatrix} 3 & 1 & -1 & 0 \\ 2 & 4 & -2 & 3 \end{pmatrix} \begin{pmatrix} 3 & 2 \\ 1 & 4 \\ -1 & -2 \\ 0 & 3 \end{pmatrix}$

   c) $\begin{pmatrix} 3 & 2 \\ 1 & 4 \\ -1 & -2 \\ 0 & 3 \end{pmatrix} \begin{pmatrix} 3 & 1 & -1 & 0 \\ 2 & 4 & -2 & 3 \end{pmatrix}$

3. Compute the inverses of the following matrices, or show that they are singular:
   a) $\begin{pmatrix} 2 & -1 \\ -2 & 3 \end{pmatrix}$

   b) $\begin{pmatrix} 1 & 0 & 1 \\ 0 & 1 & 0 \\ -1 & 0 & 1 \end{pmatrix}$ c) $\begin{pmatrix} 1 & 2 & 3 \\ 4 & 5 & 6 \\ 7 & 8 & 9 \end{pmatrix}$

3.5 Computer Projects

1. Write a program which will read in a given matrix A, and output both AA' and $A'A$.

2. Devise a program which will allow a user to input a number of matrices of various sizes, and by a set of commands or menus to multiply, sum, transpose or invert them.

CHAPTER 4

Probability Theory

4.1 Probability

Defining what is meant by probability is not easy. There are, broadly speaking, two possible approaches to the problem - the *classical* and the *Bayesian*.

The classical approach is to define the probability of an event occurring as the long-term relative proportion of times that event occurs in a series of repeated experiments. For example, if we toss a coin repeatedly, and find that in a large number of tosses we get approximately half heads and half tails, then we can say that the probability of getting heads is $\frac{1}{2}$. In general we can say that, as the number of repeated experiments increases,

$$\frac{\text{Number of times event occurs}}{\text{Number of repeated experiments}} \rightarrow \text{Probability of the event.}$$

However, it is not always possible to carry out such repeated experiments, and yet we may still want to talk about probabilities.

The Bayesian approach is to consider probability to be subjective, and that the probability of an event is the same as one's "degree of belief" in the proposition that the event will occur. For example, saying that the probability of a nuclear war this century is 0.4 is not meaningless,

though it will be subjective. Clearly such a probability
cannot be defined by the classical approach. In the coin-
tossing experiment, saying that the probability of heads is
$\frac{1}{2}$ just means that you have no reason to suppose that the coin
will land heads more often than tails, or vice versa, and
is thus a subjective judgement.

However probability is defined, there are certain
axioms of probability to which we assume it conforms. Let
us suppose that for a given experiment there is a set Ω of
elementary outcomes or *elementary events,* which are mutually
exclusive and exhaustive - i.e. one and only one of these
must occur. For coin tossing, Ω = {heads,tails}; for rolling
a die, Ω = {1,2,3,4,5,6}; for choosing Miss World, Ω is the
set of all eligible females. A general event will consist
of a subset of Ω, i.e. a set of elementary events.
Probailities are assigned to every elementary event, and the
probability of any general event is the sum of all the
probabilities of the elementary events which make it up.
The following must always be true:

1. $P(\Omega)$ = 1 .
2. $P(\emptyset)$ = 0 .
3. If $\overline{A}$ is the complement of the event A,
$P(\overline{A})$ = 1 - $P(A)$.

From these axioms it is a simple matter to derive
formulae for combinations of events. For example, $P(A \cup B)$
represents the probability that both A and B occur, while
$P(A \cap B)$ is the probability that event A or event B, or both,
occur. To compute this latter probability, consider the
Venn diagram:

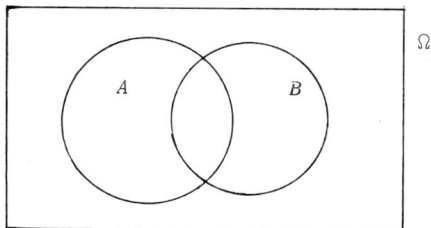

Figure 4.1

49

"Explode" the overlapping circles into three parts:

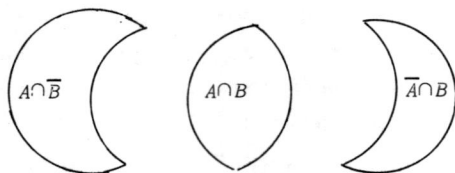

$A\cap\overline{B}$ $A\cap B$ $\overline{A}\cap B$ figure 4.2

So $P(A\cup B)$ = $P(A\cap\overline{B}) + P(A\cap B) + P(\overline{A}\cap B)$

Now A = $(A\cap\overline{B})\cup(A\cap B)$

=> $P(A)$ = $P(A\cap\overline{B}) + P(A\cap B)$

and $P(B)$ = $P(\overline{A}\cap B) + P(A\cap B)$.

Therefore, $P(A\cup B)$ = $P(A) + P(B) - P(A\cap B)$.

In the special case where A and B are mutually exclusive events, $A\cap B = \emptyset$, $P(A\cap B) = 0$, and

$$P(A\cup B) = P(A) + P(B) .$$

4.2 Conditional Probability

It is often the case that we are given partial knowledge about the outcome of an experiment, and this will revise our probabilities for the various events. For example, if we are rolling a die, and

A = Event that an even number is rolled = {2,4,6}

B = Event that a number > 3 is rolled   = {4,5,6},

then $P(A) = \frac{1}{2}$ if each elementary event has equal probability of 1/6.

If we are now told that event B has occurred, so that we know that the number is greater than 3, then the only possible elementary events are 4, 5, and 6, two of which also belong to A. So the revised probability of A is 2/3. We say the "Probability of A *given* B", or the "Probability of A *conditional on* B", which is written as $P(A|B)$, = 2/3.

To obtain a general formula for conditional probability, we should note that the only members of A that can occur once we know B has happened are those in A∩B. So the conditional probability we want is proportional to P(A∩B):

$$P(A|B) = \alpha P(A \cap B) .$$

To get a value for α, set A = B, whence

$$P(B|B) = \alpha P(B \cap B) = \alpha P(B) .$$

Now obviously P(B|B) = 1, so α = 1/P(B).
The formula is therefore:

$$P(A|B) = \frac{P(A \cap B)}{P(B)} .$$

In our die-rolling example, A∩B = {4,6}, and P(A∩B) = 1/3. P(B) = ½, so P(A|B) = (1/3)/½ = 2/3.

SPECIAL CASES

4.2.1 EXCLUSIVE EVENTS

Two events A and B are mutually exclusive if it is impossible for both to happen together; i.e. A∩B = ∅.

Therefore, $P(A \cap B) = 0$

which means $P(A|B) = 0$ and $P(B|A) = 0$.

4.2.2 INDEPENDENT EVENTS

Two events A and B are *independent* if the occurrence of one does not affect the probability of the other occurring.

i.e. $P(A|B) = P(A)$

and $P(B|A) = P(B)$.

If this is true, $P(A|B) = \dfrac{P(A \cap B)}{P(B)} = P(A)$

=> $P(A \cap B) = P(A)P(B) .$

So for independent events the probability of both occuring is

just found by multiplying together the two individual
probabilities. This is *only* true for independent events.

4.3 Combinations and Permutations

It is often the case (though not always) that we can
assume that each elementary event is equally probable. In
this case, computing probabilities just reduces to counting
elementary events. Sometimes the elementary events are ways
of arranging objects, and it is useful to have formulae for
counting such arrangements.

If n objects are to be arranged in order, in how many
different ways can this be done ? The first object in order
can be selected from the whole n, and the second from the
remaining $n-1$, and so forth until the last object can only
be chosen from the one which is left. So the total number
of arrangements is $n(n-1)(n-2) \ldots 3.2.1$. This number
is called n *factorial* and written $n!$

Now consider the problem of selecting a subset of r
objects from a larger set of n objects. In how many ways can
this be done ? The answer depends on whether or not the
order of selection is important. If the order of selection
is important, then we are dealing with *permutations* -
otherwise we are dealing with *combinations*.

4.3.1   PERMUTATIONS

To compute the number of permutations of r objects
chosen from n, consider the n objects to be arranged in
order; this can be done in $n!$ ways:

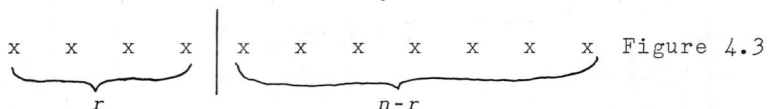

Figure 4.3

Divide off the first r objects - these form your chosen subset. However, there are not $n!$ such permutations, because we are not interested in the order of the $n-r$ "rejected" objects. These can be re-arranged in $(n-r)!$ different ways, and these all correspond to only one permutation of the r chosen objects.

So, total number of permutations $= n!/(n-r)!$
$$= n(n-1)(n-2) \ldots (n-r+1) .$$

This number is sometimes written as ^nP_r .

Example: There are 10 Miss World finalists, from whom first, second and third places are to be chosen at random. What is the probability that Miss Norway is first and Miss Nigeria third ?

The number of ways of choosing 3 in order from 10
$= {}^{10}P_3 = 10!/7! = 10.9.8 = 720.$

This is the total number of elementary events. The number of elementary events giving the desired result is equal to the number of ways in which the second place can be filled from the remaining finalists $= 8.$

So the probability is $8/720 = 1/90 .$

4.3.2 COMBINATIONS

We can follow the same argument as before to get the number of permutations of r objects from n, but this time each combination must correspond to $r!$ ways of re-arranging the chosen objects to give different permutations.

So the number of combinations $= {}^nP_r/r!$
$$= \frac{n!}{(n-r)!r!} .$$

This number is written as ^nC_r , and can be calculated by
$$^nC_r = \frac{n(n-1)(n-2) \ldots (n-r+1)}{r(r-1)(r-2) \ldots 3.2.1} .$$

4.4 The Binominal Probability Distribution

The situation which gives rise to the Binomial probability distribution is as follows: a simple experiment is repeated n times, and each repetition is known as a *trial*, and has only two possible outcomes, conventionally called "success" and "failure". The trials are all independent and have the same probability of success p. Let $q = 1-p$ be the probability of failure.

We wish to compute the probability of exactly r successes in the n trials. The total number of elementary events which give rise to the required event (r successes out of n) is equal to the number of ways of choosing the r trials which are to be successes out of the n total trials
$$= {}^n C_r \; .$$
Each such elementary event is a series of successes and failures, like
$$SSFSFFFSSFSF \ldots FSSF$$
$$(r \; S\text{'s and } n-r \; F\text{'s}).$$
Such an elementary event has probability
$$ppqpqqqppqpq \ldots qppq \; = \; p^r q^{n-r} \; .$$
So the total probability of r successes from n trials is
$$p_r \; = \; {}^n C_r \; p^r q^{n-r} \; .$$

In practice we do not normally use this formula when computing Binomial probabilities, especially by computer. We would use instead a *recurrence relationship*, which enables us to compute the value of p_{r+1} knowing the value of p_r. The reason for this is that using the formula directly requires computing $n!$, $r!$ etc., which swiftly become very large indeed and will overflow the storage capacity of any computer very soon. Dividing very large numbers by one another also leads to numerical rounding errors.

From the formula, it is clear that

$$p_r = {}^nC_r \, p^r q^{n-r} \qquad = \frac{n!}{r!(n-r)!} \, p^r q^{n-r}$$

and

$$p_{r+1} = {}^nC_{r+1} p^{r+1} q^{n-r-1} \qquad = \frac{n!}{(r+1)!(n-r-1)!} \, p^{r+1} q^{n-r-1}.$$

So the ratio

$$\frac{p_{r+1}}{p_r} = \frac{n!}{(r+1)!(n-r-1)!} \, \frac{r!(n-r)!}{n!} \, \frac{p^{r+1} q^{n-r-1}}{p^r q^{n-r}}$$

$$= \left(\frac{n-r}{r+1}\right)\left(\frac{p}{q}\right) \text{, after cancelling.}$$

Thus,

$$p_{r+1} = \left(\frac{n-r}{r+1}\right)\left(\frac{p}{q}\right) p_r .$$

To start the calculations, we just note that

$$p_0 = q^n .$$

Then

$$p_1 = \frac{n}{1} \frac{p}{q} p_0 \text{ etc.}$$

As well as the probabilities $\{p_r\}$, it is possible to compute the *cumulative probabilities* $\{F_r\}$, where

$$F_r = P(\text{No. of successes} \le r)$$

$$= \sum_{i=0}^{n} p_i .$$

Again, it is easiest to use a recurrence relationship:

$$F_0 = p_0 = q^n ,$$
$$F_{r+1} = F_r + p_{r+1} .$$

Using these relationships, it is easy to write a program to compute and print out Binomial probabilities and cumulative probabilities.

```
program binomial(input,output);
{ Program to print out Binomial probabilities }
var n,r,i,j : integer;
    prob,cum,p,q : real;
    ans : char;
begin
  writeln; write('Input no. of trials: '); read(n);
  writeln; write('Input success probability: '); read(p);
  writeln; write('Do you want just a single value (y or n)?');
  readln; read(ans); writeln;
```

```
if ans = 'y' then
begin
  write('Input required no. of successes: '); read(r);
  writeln;
end
else r := -1;
q := 1.0 - p; i := 0; prob := 1.0;
for j := 1 to n do prob := prob*q;
writeln;
writeln('No. of successes    Prob.    Cum. Prob.');
writeln;
cum := prob;
if (r = -1) or (r = 0) then
  writeln(i:8,prob:16,cum:16);
while (i <> r) and (i < n) do
begin
  prob := (n-i)*p*prob/((i+1)*q);
  cum := cum + prob;
  i := i + 1;
  if (i = r) or (r = -1) then
    writeln(i:8,prob:16,cum:16);
end;
end.
```

4.5 Probability Distributions

A *random variable* is the unknown numerical outcome of
an experiment - for example, the number of successes in n
Binomial trials. If X is a random variable with N possible
values $x_1, x_2, \ldots x_N$ (in ascending order), then the
probability distribution for X is the set of corresponding
probabilities $p_1, p_2, \ldots p_N$, with

$$P(X=x_i) \quad = \quad p_i \ , \ i = 1, \ldots N$$

and
$$\sum_{i=1}^{N} p_i \quad = \quad 1 \ .$$

We may also define the cumulative probabilities:

$$F_i \quad = \quad P(X \le x_i) \quad = \quad \sum_{j=1}^{i} p_j \ .$$

Note:
$$F_N \quad = \quad P(X \le x_N) \quad = \quad 1 \text{ always.}$$

Often there is a formula which enables us to compute
these probabilities for a given distribution. For example,
if X is a Binomial random variable, then we have produced a

56

formula for the probability distribution in section 4.4.

Even if we do have such a formula, it is often helpful to be able to summarise the behaviour of the random variable X by a few relevant numbers rather than a whole set of probabilities. One important such number is the *mean*, which gives a measure of the "average" value of X. Its definition is:

$$\text{Mean} \quad \mu \quad = \quad \sum_{i=1}^{N} p_i x_i \; .$$

For example, if X is the number of heads obtained when a coin is tossed 4 times, then the values of the probability distribution for X are obtained from the Binomial formula with $n = 4$ and $p = \frac{1}{2}$.

x_i:	0	1	2	3	4
p_i:	$\frac{1}{16}$	$\frac{1}{4}$	$\frac{3}{8}$	$\frac{1}{4}$	$\frac{1}{16}$

$$\text{Mean} \quad \mu = 0\times\frac{1}{16} + 1\times\frac{1}{4} + 2\times\frac{3}{8} + 3\times\frac{1}{4} + 4\times\frac{1}{16} = 2.$$

This corresponds to our intuitive feeling that on average we would expect to get two heads from 4 tosses. However, the mean is not always going to be an integer, even if the values of X are all integers. Consider the case when X is the number obtained by rolling a fair die:

x_i:	1	2	3	4	5	6
p_i:	$\frac{1}{6}$	$\frac{1}{6}$	$\frac{1}{6}$	$\frac{1}{6}$	$\frac{1}{6}$	$\frac{1}{6}$

$$\text{Mean} \quad \mu = (1+2+3+4+5+6)\times\frac{1}{6} = 3\tfrac{1}{2} \; .$$

As well as the mean μ it is useful to have a measure of the variability or dispersion of X - the amount by which it varies either side of the mean. To do this, we define a set of *deviations* from the mean, such that the ith deviation is

$$d_i \quad = \quad x_i \quad - \quad \mu.$$

The *variance* σ^2 is defined to be the mean of the squares of these deviations:

$$\sigma^2 \quad = \quad \sum_{i=1}^{N} p_i d_i^2 \quad = \quad \sum_{i=1}^{N} p_i (x_i - \mu)^2 \; .$$

The *standard deviation* σ is the square root of the variance, and will thus be in the same units as X itself. In some situations this will be a more useful measure than the variance, but in others we shall be using the variance.

For the example with X = number of heads in 4 tosses:

$$d_i = x_i - \mu: \qquad -2 \qquad -1 \qquad 0 \qquad 1 \qquad 2$$
$$d_i{}^2: \qquad 4 \qquad 1 \qquad 0 \qquad 1 \qquad 4$$
$$p_i: \qquad \frac{1}{16} \qquad \frac{1}{4} \qquad \frac{3}{8} \qquad \frac{1}{4} \qquad \frac{1}{16}$$

Variance $\sigma^2 = 4 \times \frac{1}{16} + 1 \times \frac{1}{4} + 0 \times \frac{3}{8} + 1 \times \frac{1}{4} + 4 \times \frac{1}{16} = 1$.

So the standard deviation $\sigma = 1$ also.

For the die-rolling experiment, with $\mu = 3\frac{1}{2}$:

$$d_i: \qquad -2\frac{1}{2} \qquad -1\frac{1}{2} \qquad -\frac{1}{2} \qquad \frac{1}{2} \qquad 1\frac{1}{2} \qquad 2\frac{1}{2}$$
$$d_i{}^2: \qquad \frac{25}{4} \qquad \frac{9}{4} \qquad \frac{1}{4} \qquad \frac{1}{4} \qquad \frac{9}{4} \qquad \frac{25}{4}$$

$$\sigma^2 = \sum_{i=1}^{6} p_i d_i^2 = \frac{70}{24} = 2.9167 .$$

Standard deviation $\sigma = \sqrt{2.9167} = 1.7078$.

Another way of expressing all this is by means of the *expectation operator* E(). This transforms a random variable, or a function of a random variable, into a real number in the following way:

$$E(X) = \sum_{i=1}^{N} p_i x_i .$$

For any function of X:

$$E(f(X)) = \sum_{i=1}^{N} p_i f(x_i) .$$

So

$$E(X^2) = \sum_{i=1}^{N} p_i x_i^2 ,$$

$$E(\ln(X)) = \sum_{i=1}^{N} p_i \ln(x_i) \qquad \text{etc.}$$

Certain rules clearly hold when using the expectation operator:

1. $E(X+a) = E(X) + a$ (a is constant) .
2. $E(aX) = aE(X)$.
3. E() does not interchange with other operators - for example:
$$E(X^2) \neq (E(X))^2 ,$$
$$E(\ln(X)) \neq \ln(E(X)) \qquad \text{etc.}$$
4. We can define the mean and variance in terms of the

the expectation operator:

$$\mu = E(X) ,$$
$$\sigma^2 = E((X-\mu)^2) .$$

Often it is possible to derive *expectation formulae* for the mean and variance of particular probability distributions. For example, for a Binomial random variable X with n trials and success probability p, we have the formula for the probability that $X = i$:

$$p_i = {}^nC_i \, p^i q^{n-i} = \frac{n!}{i!(n-i)!} \, p^i q^{n-i} .$$

So
$$\mu = E(X) = \sum_{i=0}^{n} p_i \, i$$
$$= \sum_{i=0}^{n} \frac{n!}{i!(n-i)!} \, p^i q^{n-i} \, i .$$

Now, in this expression the term for $i = 0$ is zero, so

$$\mu = \sum_{i=1}^{n} \frac{n!}{(i-1)!(n-i)!} \, p^i q^{n-i}$$
$$= np \sum_{i=1}^{n} \frac{(n-1)!}{(i-1)!(n-i)!} \, p^{i-1} q^{n-i} .$$

Now set $j = i-1$ and $m = n-1$, and we get:

$$\mu = np \sum_{j=0}^{m} \frac{m!}{j!(m-j)!} \, p^j q^{m-j} .$$

The sum is just a sum of Binomial probabilities with m trials, and must equal 1.

So
$$\underline{\mu = E(X) = np} .$$

There is a similar formula for the variance of a Binomial probability distribution, which takes longer to derive:

$$\underline{\sigma^2 = Var(X) = npq} .$$

There are several other useful probability distributions which are related to the Binomial in various ways.

4.5.1 THE POISSON DISTRIBUTION

This is the limiting case of the Binomial distribution as the number of trials tends to ∞. If we set $\lambda = np =$ mean

59

number of successes, and allow n to tend to ∞ and p to tend to
0 in such a way that λ stays constant, then the result is that
the random variable X (number of successes) has a *Poisson*
distribution, with probabilities:

$$p_i \;=\; P(X{=}i) \;=\; \frac{\lambda^i e^{-\lambda}}{i!} \quad \text{for } i = 0,1,2,3 \ldots$$

The mean and variance of this distribution are:

$$E(X) \;=\; \lambda \quad \text{and} \quad Var(X) \;=\; \lambda \text{ also.}$$

Applications of the Poisson distribution are in
situations where a large number of possible trials each have
a very low probability of success. For example, a complex
system contains a large number of components each with a low
failure probability. On average, one component fails per
hour. What is the probability that at least 2 will fail in a
given hour ?

Computing the Poisson probabilities with $\lambda = 1.0$:

i	p_i	$F_i = P(X{\leq}i)$
0	0.3679	0.3679
1	0.3679	0.7358
2	0.1840	0.9198

$$P(X{\geq}2) \;=\; 1 - P(X{<}2) \;=\; 1 - F_1 \;=\; 1 - 0.7358$$
$$=\; 0.2642 \;.$$

A recurrence relationship can be set up to compute
these probabilities:

$$p_i \;=\; \frac{\lambda^i e^{-\lambda}}{i!} \quad , \qquad p_{i+1} \;=\; \frac{\lambda^{i+1} e^{-\lambda}}{(i+1)!} \quad .$$

So
$$p_{i+1} \;=\; \frac{\lambda}{i+1} \, p_i \;.$$

4.5.2   THE NEGATIVE BINOMIAL DISTRIBUTION

In this case we have independent trials each with
probability p of success, and continue the trials until
exactly r successes are obtained. The random variable is n,
the number of trials needed to obtain the specified number of

successes.

$$p_n = P(n \text{ trials to get exactly } r \text{ successes})$$
$$= P(n\text{th trial a success}) \times P(r-1 \text{ successes in } n-1)$$
$$= p \times {}^{n-1}C_{r-1} \, p^{r-1} q^{n-r}$$
$$= {}^{n-1}C_{r-1} \, p^r q^{n-r} \quad \text{for} \quad n = r, r+1, r+2, \ldots$$

The mean $E(n) = r/p$.

A recurrence relationship for the probabilities is:

$$p_{n+1} = \frac{n}{n-r+1} q \, p_n \; .$$

4.5.3 THE GEOMETRIC DISTRIBUTION

In this case, n is the number of trials to the first success. It is a special case of the Negative Binomial distribution with $r = 1$.

$$p_n = P(n \text{ trials to get first success}) = pq^{n-1} \; .$$

These probabilities form a geometric progression - each is q times the one before.

The mean $E(n) = 1/p$.

Recurrence relationship for probabilities:

$$p_{n+1} = q \, p_n \; .$$

4.6 Continuous Probability Distributions

So far all our random variables have only taken on discrete values - either a finite number (such as the Binomial), or a countably infinite number (such as the Poisson). However, it is commonly the case that we are dealing with a continuous range of possible values for a random variable X.

For example, consider a "Wheel of Fortune", whose pointer can be spun, and let X be the angle in degrees

61

between the pointer and the vertical when it comes to rest.

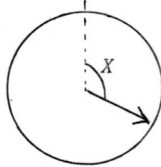

Figure 4.4

So X takes all values between 0 and 360 with equal probability, which means that

$$P(X=90) \quad = \quad 1/N \ ,$$

where N is the total number of possible values between 0 and 360. But $N = \infty$, which means that $P(X=90) \ = \ 0$.

This argument is true for any value of X, so the probability that X exactly equals any precise value is 0. But if we actually carry out the experiment and measure X precisely, we will find it equals some value, x degrees say. Now beforehand, $P(X=x) = 0$, so an event with probability zero has occurred.

When dealing with continuous random variables, thinking about such probabilities can cause brain-ache, so we shall make a rule never to discuss the probability that the random variable exactly equals some value, but only the probability that it is in some interval; for example,

$P(X$ is in the interval $(89.5,90.5))$ is clearly equal to $1.0/360.0 \ = \ $ (Width of interval)/(Total width) .

To discuss continuous random variables, we shall define for each such variable two functions which describe its behaviour. We can no longer use the old description of quoting the probabilities for the different values of X, as these are all zero.

The first descriptive function is the *Probability Distribution Function* for X, written as $F()$. It is defined by:

$$F(x) \quad = \quad P(X \in (-\infty, x)) \quad = \quad P(X \leq x) \ .$$

We can use this to compute the probability that X lies in any interval:

$$P(X \in (x_1, x_2)) \quad = \quad P(X \leq x_2) \ - \ P(X \leq x_1)$$

$$= \quad F(x_2) \ - \ F(x_1) \ .$$

(*Note:* It does not matter whether we write $P(X \leq x)$ or $P(X < x)$
     - these are equal, as $P(X=x) \ = \ 0$).

We may plot $F(x)$ as a function of x, to get a graphical representation of the distribution of X:

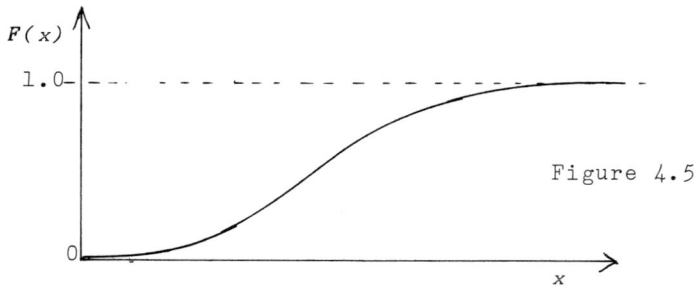

Figure 4.5

The function $F()$ has the following properties:
1. It is a monotonically non-decreasing function of x:
   if $x_2 > x_1$, then $F(x_2) \geq F(x_1)$.
2. $F(\infty) = P(X \leq \infty) = 1$.
3. $F(-\infty) = P(X \leq -\infty) = 0$.

The second descriptive function gives a better overall impression of the shape of the distribution of X - it attempts to answer the question "What is the probability that X is 'round about' the value x ?".

The *Probability Density Function* $f()$ is not a probability; it is defined as follows:

Take a small change δx in x; then
$$P(X \in (x, x+\delta x)) \approx f(x) \, \delta x .$$
$f(x)$ thus measures the "density" of the probability at the value x. Plotting this function gives a picture like:

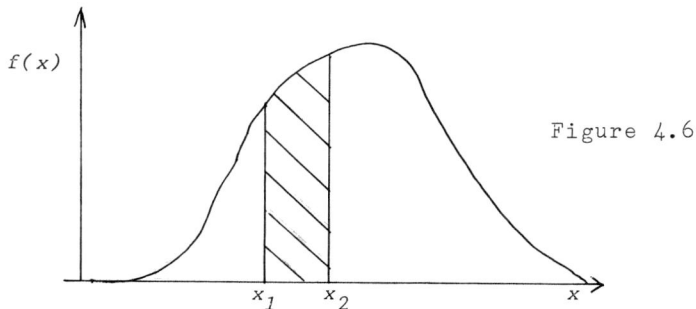

Figure 4.6

$P(X \in (x_1, x_2)) = $ Area under the curve between x_1 and x_2

$$= \int_{x_1}^{x_2} f(x) \, dx \; .$$

So $\quad P(X \leq x) \quad = \quad \int_{-\infty}^{x} f(x) \, dx \quad = \quad F(x) \, .$

This shows the relationship between the two functions $F()$ and $f()$:

$$F(x) \quad = \quad \int_{-\infty}^{x} f(x) \, dx$$

and $\quad f(x) \quad = \quad \dfrac{dF(x)}{dx} \quad = \quad F'(x) \; .$

The probability density function $f()$ has the following properties:

1. $f(x) \geq 0$ for all x .

2. $\int_{-\infty}^{\infty} f(x) \, dx \quad = \quad 1.$

There are a large number of continuous probability distributions which are commonly used - two examples now follow.

4.6.1 UNIFORM DISTRIBUTION

If X is uniformly distributed between 0 and a, then its probability density function has the following form:

$$f(x) \quad = \quad b \quad 0 \leq x \leq a$$
$$\quad = \quad 0 \quad \text{otherwise} \; .$$

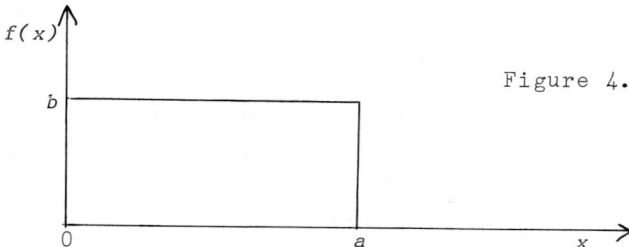

Figure 4.7

To find the value of b, note that the total area $= ab$, and this must be equal to 1, which means that $b = 1/a$.

$$F(x) \quad = \quad \int_{0}^{x} f(x) \, dx \quad = \quad \int_{0}^{x} \dfrac{dx}{a} \quad = \quad \dfrac{x}{a} \quad (0 \leq x \leq a).$$

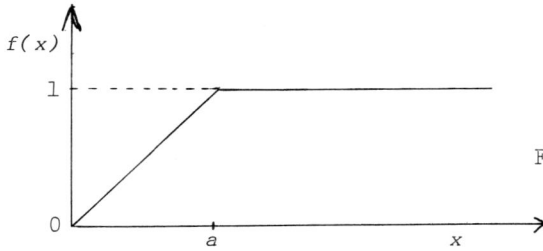

Figure 4.8

4.6.2 NEGATIVE EXPONENTIAL DISTRIBUTION

$$f(x) \quad = \quad \lambda e^{-\lambda x} \qquad x \geq 0$$
$$= \quad 0 \qquad\quad x < 0 .$$

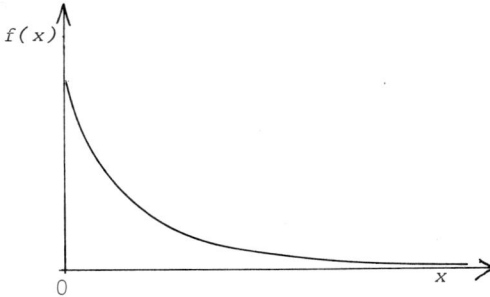

Figure 4.9

$$F(x) \quad = \quad \int_0^x f(x)\ dx \quad = \quad \int_0^x \lambda e^{-\lambda x}\ dx$$

$$= \quad \left[-e^{-\lambda x} \right]_0^x \quad = \quad \underline{1 - e^{-\lambda x}} .$$

As x tends to ∞, $F(x)$ tends to 1.

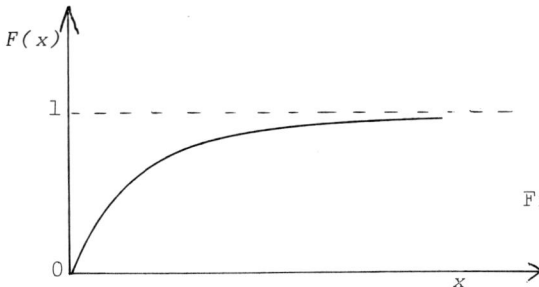

Figure 4.10

65

We can no longer use the same definition of expectation as for a discrete random variable. We write instead, if $h(X)$ is a function of X:

$$E(h(X)) \quad = \quad \int_{-\infty}^{\infty} h(x)\ f(x)\ dx \quad .$$

So $\quad E(X) \quad = \quad \mu \quad = \quad \int_{-\infty}^{\infty} x\ f(x)\ dx \quad .$

$$\text{Var}(X) \quad = \quad E((X-\mu)^2) \quad = \quad \int_{-\infty}^{\infty} (x-\mu)^2\ f(x)\ dx \ .$$

Examples: 1. Uniform distribution.

$$E(X) \quad = \quad \int_0^a \frac{x}{a}\ dx \quad = \quad \left[\frac{x^2}{2a} \right]_0^a \quad = \quad \frac{a}{2} \quad .$$

$$\text{Var}(X) \quad = \quad \int_0^a \left(x-\frac{a}{2}\right)^2 \frac{dx}{a}$$

$$= \quad \int_0^a \left(x^2 - ax + \frac{a^2}{4}\right) \frac{dx}{a}$$

$$= \quad \frac{1}{a} \left[\frac{x^3}{3} - \frac{ax^2}{2} + \frac{a^2 x}{4} \right]_0^a \quad = \quad \frac{a^2}{12} \quad .$$

2. Negative exponential distribution.

$$E(X) \quad = \quad \int_0^{\infty} x\lambda e^{-\lambda x}\ dx \quad = \quad \frac{1}{\lambda} \quad .$$

4.7 Sums and Differences of Random Variables

Suppose X and Y are two independent random variables with means μ_x and μ_y, and variances σ_x^2 and σ_y^2, and that we define a new random variable $Z = X + Y$. What are the mean and variance of Z ?

$$E(Z) \quad = \quad E(X+Y) \quad = \quad E(X) + E(Y) \quad = \quad \mu_x + \mu_y \quad .$$

$$\begin{aligned}
\mathrm{Var}(Z) &= E((X+Y-\mu_x-\mu_y)^2) \\
&= E(X^2+2XY+Y^2-2X\mu_x-2Y\mu_y-2X\mu_y-2Y\mu_x+\mu_x^2+2\mu_x\mu_y+\mu_y^2) \\
&= E(X^2-2X\mu_x+\mu_x^2) + E(Y^2-2Y\mu_y+\mu_y^2) + 2E(XY) - 2\mu_x\mu_y \,.
\end{aligned}$$

If X and Y are independent, then $E(XY) = \mu_x\mu_y$, so

$$\begin{aligned}
\mathrm{Var}(Z) &= E((X-\mu_x)^2) + E((Y-\mu_y)^2) \\
&= \underline{\sigma_x^2 + \sigma_y^2} \,.
\end{aligned}$$

So, to get the mean and variance of the sum of two independent random variables, we must add the means and *variances* of the individual variables.

If $Z = X - Y$, we can follow almost exactly the same argument to show that:

$$E(Z) = \mu_x - \mu_y \,,$$

and
$$\mathrm{Var}(Z) = \sigma_x^2 + \sigma_y^2 \,.$$

Note that we still add the variances, even when Z is the difference between the random variables rather than their sum. We can generalise this to any number of independent variables. If $Z = X_1 \pm X_2 \pm \ldots \pm X_n$, where the means are $\mu_1, \mu_2, \ldots \mu_n$ and the variances are $\sigma_1^2, \sigma_2^2, \ldots \sigma_n^2$, then

$$E(Z) = \mu_1 \pm \mu_2 \pm \ldots \pm \mu_n \,,$$

and
$$\mathrm{Var}(Z) = \sigma_1^2 + \sigma_2^2 + \ldots + \sigma_n^2 \,.$$

A special case is when all the μ_i and σ_i^2 values are the same, and

$$Z = \sum_{i=1}^{n} X_i \,.$$

Then
$$E(Z) = n\mu$$
and
$$\mathrm{Var}(Z) = n\sigma^2 \,.$$

If we write $\overline{X}$ as the average of the n variables:

$$\overline{X} = \frac{1}{n} \sum_{i=1}^{n} X_i = Z/n \,,$$

then
$$E(\overline{X}) = E(Z)/n = \mu \,,$$

and
$$\begin{aligned}
\mathrm{Var}(\overline{X}) &= \mathrm{Var}(Z/n) = \mathrm{Var}(z)/n^2 \\
&= n\sigma^2/n^2 = \underline{\frac{\sigma^2}{n}}
\end{aligned}$$

An important result concerns the way in which the probability distribution of $\overline{X}$ behaves as *n* increases. The *Central Limit Theorem* says that, if the random variables $\{X_i\}$ are idependent and identically distributed, then, as *n* tends to infinity, $\overline{X}$ tends to have so-called *Normal* probability distribution with mean μ and variance σ^2/n.

Because of this theorem, the Normal probability distribution is extremely important in Statistics.

4.8 The Normal Distribution

If a random variable *X* is Normally distributed with mean μ and variance σ^2, then its probability density function is:

$$f(x) \quad = \quad \frac{1}{\sqrt{2\pi}\sigma} \, exp\left[\, -\frac{(x-\mu)^2}{2\sigma^2} \, \right] \, .$$

Graphically, this looks like:

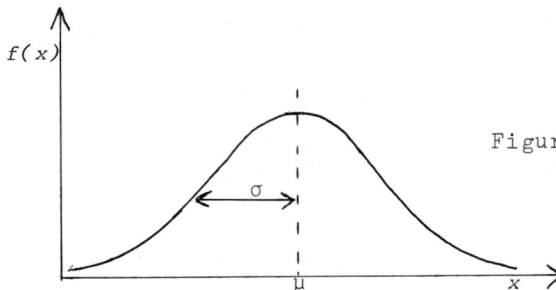

Figure 4.11

Unfortunately, there is no algebraic function which is the anti-derivative of *f()*, so it is not possible to produce an exact formula for the probability distribution function *F()*. This makes the Normal distribution slightly inconvenient, despite it being such a commonly-encountered distribution. To obtain $F(x)$, we will need to use some technique of numerical integration, such as Simpson's Rule (Chapter 2). However, this becomes very tedious if we are going to use the Normal distribution a lot. We make use of a simple transformation which allows us to look up a reasonably small set of tabulated values.

Define a random variable $Z = (X-\mu)/\sigma$. Then Z is also Normally distributed, with mean 0 and variance 1, and is called a *Standard Normal* random variable. Its density function is

$$f(z) = \frac{1}{\sqrt{2\pi}} e^{-z^2/2} \ .$$

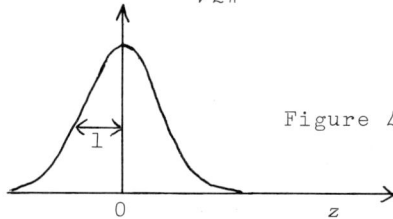

Figure 4.12

Its distribution function is called the *Standard Normal Integral Function*, and is defined as:

$$\Phi(z) = P(Z \leq z) = \frac{1}{\sqrt{2\pi}} \int_{-\infty}^{z} e^{-z^2/2} \, dz \ .$$

Tables of this function can be consulted for any value of z.

So the basic strategy for computing probabilities for any Normal random variable X is to transform the problem to an equivalent one in the Standard Normal variable Z and use the tables of Φ.

Example: X is Normally distributed with mean 10 and variance 4. What is the probability that X is less than 8 ?

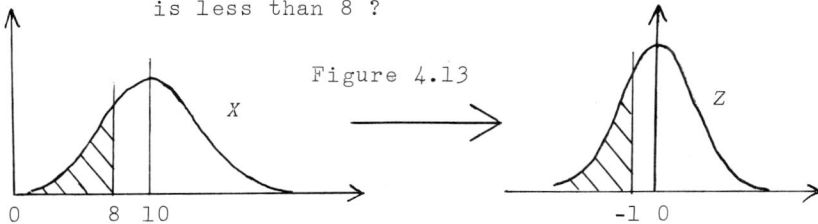

Figure 4.13

69

$$\text{Transform} \quad Z \quad = \quad \frac{X-\mu}{\sigma} \quad = \quad \frac{X-10}{2} \; .$$

$$\text{So} \quad P(X<8) \quad = \quad P(Z < \tfrac{8-10}{2}) \quad = \quad P(Z < -1)$$

$$= \quad \Phi(-1) \quad = \quad \underline{0.1587} \quad \text{from tables.}$$

4.8.1 USE OF THE STANDARD NORMAL TABLES

In Table 1, the values given are not actually $\Phi(z)$, but $P(0 \le Z \le z) \quad = \quad \Phi(z) - 0.5$.

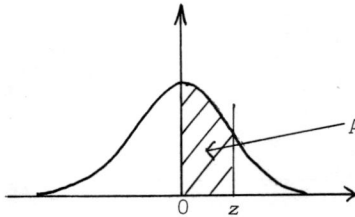

Figure 4.14

Area given in Table 1.

To use this table, we must just remember that the Standard Normal distribution is symmetrical about 0. This means:

Figure
4.15

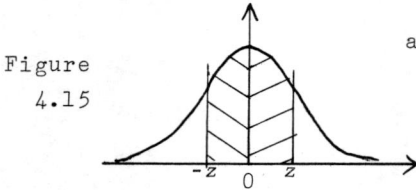

a) $P(-z \le Z \le 0)$
$= \quad P(0 \le Z \le z)$.

Figure
4.16

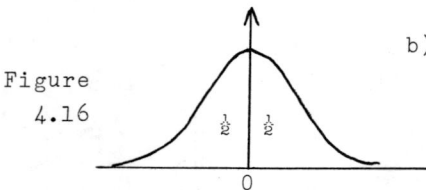

b) $P(-\infty \le Z \le 0)$
$= \quad P(0 \le Z \le \infty) \quad = \quad \tfrac{1}{2}$.

Example: X has mean 12 and standard deviation 3. What is the probability that X lies between 11 and 14 ?

$$\text{Transform} \quad Z \quad = \quad \frac{X-12}{3} \; .$$

So $P(11 \le X \le 14) = P(-\frac{1}{3} \le Z \le \frac{2}{3})$.

Figure 4.17

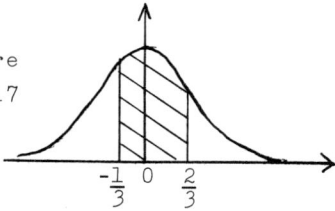

The area from 0 to 2/3 is obtained directly from Table 1 with $z = 0.667$, and equals 0.2475.
The area from -1/3 to 0 is obtained with $z = 0.333$, and equals 0.1306.

Therefore, total probability = 0.3781 .

Note that sums, differences and means of Normal random variables are themselves also Normal random variables.

Example: A random sample of 10 values is taken from a Normal random variable with mean 20 and variance 40. What is the probability that the average of the sample will be less than 16 ?

The average $\bar{X}$ is Normally distributed with mean 20 and variance 40/10 = 4.0 .

So $Z = \frac{\bar{X}-20}{2}$ and

$$P(\bar{X} < 16) = P(Z < -2)$$
$$= 0.5 - \text{Table probability}$$
$$= 0.5 - 0.4772$$
$$= 0.0228 .$$

Figure 4.18

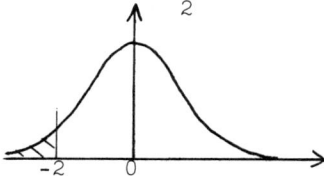

4.9 Exercises

1. A card is drawn at random from a pack of 52. Let A be the event that the card is a spade, and B be the event that it is a black card higher than the 10 (Aces high). Find:

71

a) $P(A)$ b) $P(B)$ c) $P(A \cap B)$ d) $P(A \cup B)$

e) $P(A|B)$ f) $P(B|A)$.

2. Fred and Bert play a simple game, as follows: a coin is
    tossed and a die is rolled. Fred wins if the coin lands
    heads and the die score is even, or if the coin lands
    tails and the die score is odd. Otherwise Bert wins.
    Fred has smuggled a loaded die into the game, which has
    probabilities 1/8, 1/6, 1/6, 1/6, 1/8 and 1/4 for the
    numbers 1 to 6 respectively. Not to be outdone, Bert is
    using a trick penny which lands tails twice as often as
    it lands heads.
    a) What is the probability that Fred wins ?
    b) What is the probability that Bert wins ?
    c) If the coin lands heads, what is the probability of
       Fred winning ?
    d) If the number on the die is greater than 3, what is the
       probability of Bert winning ?
    e) Would Fred be better off not trying to cheat ?

3. A student at Dr. Turtle's School for Porpoises has a
    probability of 0.12 of failing his Distraction exam, a
    probability of 0.29 of failing in Mystification, and the
    probability that he fails both is 0.07 .
    a) What is the probability that he fails at least one
        exam ?
    b) What is the probability that he fails just one exam ?
    c) What is the probability that he passes both exams ?
    d) What is the probability that he passes in Distraction,
        conditional on him passing in Mystification ?

4. In how many different ways can 5 cards be chosen from a
    pack of 52 ? What is the probability that these 5 cards
    contain 4 aces ? What is the probability that they form
    a flush (all 5 cards the same suit) ?

5. An urn contains one black ball, two white balls, three
    blue balls and four red balls. Two balls are drawn from
    the urn without being replaced. What are the probabilities

of the following events ?
a) Both balls are white.
b) Both balls are blue.
c) Both balls are red.
d) One black ball and one blue ball.
e) One white ball and one red ball.
f) At least one red ball.

6. If on average rain falls on 12 days out of 30, find the
   probabilities of the following:
   a) The first 3 days of a week are fine and the rest of
      the week is wet.
   b) Rain falls on exactly 3 days in a week.
   c) Rain falls on no more than 3 days in a week.

7. A chain is made from 10 links, and the breaking load for
   each link is uniformly distributed between 50 and 60 lbs.
   What is the probability that the chain will break when
   subjected to a load of 51 lbs. ?

8. In a Computer Studies Department, terminals are allocated
   for the use of a class of 20 students, who need them for
   10% of their working time.
   a) If 3 terminals are allocated, what is the probability
      that there will be more students wanting terminals
      than terminals available at any arbitrary time ?
   b) How many terminals must be allocated to reduce the
      probability of an excess of students over terminals
      to below 5% ?

9. In an engineering workshop each man needs to use a
   polishing machine for 30% of his working time.
   a) How many machines are needed for 10 men, if demand
      for them is to be met at least 95% of the time ?
   b) How many machines are needed for a similar service to
      60 men ?
   (Hint: use a computer program) .

10. A bag contains 4 black balls and 4 white balls. 4 balls are drawn at random from the bag, and not replaced. If X is the number of black balls drawn, find:
    a) The probability distribution of X.
    b) The mean of X.
    c) The standard deviation of X.
    d) $E(1/(1+X))$.

11. On average two golfers are struck by lightning each year. Use a suitable probability distribution to calculate the probability of more than three golfers being struck next year.

12. A potter makes teapots, 25% of which crack when being fired. He makes them one at a time, and has an order for 8. What is the probability that he will have to make more than 10 teapots to get the 8 he needs ?

13. A continuous random variable X has probability density function:
$$f(x) \quad = \quad 0 \quad \text{for} \quad x < 1 \ ,$$
$$\quad = \quad \frac{1}{x^2} \quad \text{for} \quad x \geq 1 \ .$$
Find an expression for $F(x)$, and the probability that X is greater than 3. What is strange about the mean of this distribution ?

14. A continuous random variable Y has probability density function:
$$f(y) \quad = \quad 0 \quad \text{for} \quad y < 0 \ ,$$
$$\quad = \quad 4by - by^2 \quad \text{for} \quad 0 \leq y \leq 4 \ ,$$
$$\quad = \quad 0 \quad \text{for} \quad y > 4 \ .$$
Calculate the value of b and the mean of Y.

15. X is Normally distributed with mean 100 and standard deviation 20. What is the probability that X is greater than 90 ?

16. A machine puts shafts into bearings randomly selected from

74

two large bins. The external diameters of the shafts and
the internal diameters of the bearings are Normally
distributed with the following means and variances:

Dimension	Mean (ins)	Variance.(in^2)
External diameter	2.031	4.29×10^{-6}
Internal diameter	2.035	1.96×10^{-6}

a) What proportion of shafts will not fit into the
   bearings ?

b) What proportion of the pairs will have a clearance of
   0.002 to 0.007 inches between shaft and bearing ?

17. An examiner knows from experience that the marks obtained
on his exam are Normally distributed with mean 55% and
atandard deviation 10%. He wishes to set boundaries for
the various exam grades, so as to obtain on average the
following percentages in the various grades:

Class 1	:	5%
Class 2, division 1	:	20%
Class 2, division 2	:	35%
Class 3	:	20%
Pass	:	10%
Fail	:	10% .

What marks should he set as the class boundaries ?

18. A relay team consists of four athletes who each run one
lap. During training they have been extensively timed
over single laps, and all the times are Normally
distributed with means and standard deviations as shown:

Athlete	Mean time	Standard deviation
A	1 min 0.8 secs	2 secs
B	1 min 1.3 secs	2 secs
C	1 min 2.1 secs	4 secs
D	1 min 1.8 secs	5 secs

a) If the record for the event is 4 mins 2.8 secs, what
   is the probability of the team breaking the record ?

b) What is the probability that D will run faster than B ?

4.10 Computer Projects

1. A roulette wheel has equal divisions from 0 to 36. It is
    possible to bet on various sets of numbers, for example:

    E = {Even numbers, excluding 0},
    G = {Numbers > 18},
    P = {Prime numbers (including 1, excluding 0)} ,
    S = {Perfect squares (excluding 0)} .

    Write a program which will accept as data any two
    arbitrary sets of numbers A and B, and output the
    following probabilities:
    a) P(A)   b) P(B)   c) P($A \cap B$)   d) P($A \cup B$)
    e) P($A | B$) .
    Test your program using E and G, G and P, E and S etc.
    from the above examples.

2. Write a program which will compute probabilities for
    various discrete distributions, under the user's control.
    It should be able to compute Binomial, Poisson, Negative
    Binomial and Geometric probabilities. (Note - use
    recurrence relationships).

3. Write a program to tabulate $\Phi(z)$ for various values of z,
    using Simpson's Rule. Compare your results with Table 1.
    How many sub-intervals do you need to get 4-figure
    accuracy ?

Suggested Further Reading for Section I

"The Intensive Pascal Course", by Mick Farmer, Chartwell-
    Bratt Ltd., 1984.

"A Practical Introduction to Pascal", by I.R. Wilson and
    A.M. Addyman, MacMillan, 1978.

"Introduction to Pascal", by Jim Welsh and John Elder,
    Prentice-Hall, 1979.

"Computer Mathematics" by D.J. Cooke and H.E. Bez, Cambridge
    University Press, 1984.

"Finite Mathematics", by Daniel P. Maki and Maynard Thompson,
    McGraw-Hill, 1983.

"Calculus and Analytic Geometry", by Philip Gillett, D.C.
    Heath & Co., 1981.

"Probability with Statistical Applications", by Frederick
    Mosteller, Robert E.K. Rourke, and George B. Thomas,
    Addison-Wesley, 1970.

SECTION II

Statistics

CHAPTER 5

Estimation

Much of this book is going to be about *models* of one sort or another. A model is an imaginary entity, usually expressed in mathematical form, which we use to represent some aspect of the real world. With it we hope to understand the behaviour of the real system more fully, and from it we wish to deduce practical consequences.

Some models represent active, dynamic processes - for example, population growth or jobs queuing in a computer system. Others represent the way in which numerical data is distributed, and can be used to make statements about what may or may not be true in the real world. In this latter case, the model often consists of a probability distribution with one or more parameters, whose exact values are unknown. Before the model can be used, we need to find values for these parameters which best fit some known data. This is the problem of parameter estimation.

Example: A random variable X is assumed to be Normally distributed with unknown mean μ. N independent values of the variable have been generated: $x_1, x_2, \ldots, x_N$. Estimate μ.

We shall use the notation ^ to represent an estimated value - so the problem is one of finding $\hat{\mu}$. For example:

1. Set $\hat{\mu}_1 = x_1$.

2. If $x_{(1)}$ is the lowest of the data values, and $x_{(N)}$ is the highest value, set $\hat{\mu}_2 = \frac{1}{2}(x_{(1)} + x_{(N)})$.

3. Set $\hat{\mu}_3$ = $\overline{x}$ = $\frac{1}{N} \sum\limits_{i=1}^{N} x_i$.

4. Set $\hat{\mu}_4$ = $\frac{N+1}{N} \overline{x}$.

Now, which of these estimators is the best one to use ? We need some criteria for deciding which methods of estimating unknown parameters are useful and which are less useful.

5.1 Desirable Properties of Estimators

Suppose θ is an unknown parameter whose value we wish to estimate, and $\hat{\theta}$ is an estimated value, calculated from the data $x_1, x_2, \ldots, x_N$ according to some rule. There are certain properties which it is desirable for $\hat{\theta}$ to have.

5.1.1 UNBIASEDNESS

$\hat{\theta}$ is a random variable, since it depends on the randomly produced data $x_1, x_2, \ldots x_N$. It is *unbiased* if
$$E(\hat{\theta}) = \theta.$$
In other words, if the estimated value $\hat{\theta}$ is equal to the true value θ "on average". Turning to our Normal example, we can test each estimator to see if it is unbiased.

$$E(\hat{\mu}_1) = E(x_1) = \mu \text{ (unbiased)} .$$

$$E(\hat{\mu}_2) = \tfrac{1}{2}E(x_{(1)}) + \tfrac{1}{2}E(x_{(N)})$$
$$= \mu \text{ (unbiased)} .$$

$$E(\hat{\mu}_3) = \tfrac{1}{N} N\mu = \mu \text{ (unbiased)} .$$

$$E(\hat{\mu}_4) = \tfrac{N+1}{N} \mu \neq \mu \text{ (biased)} .$$

5.1.2 CONSISTENCY

$\hat{\theta}$ is a *consistent* estimator of θ if, as $N \to \infty$, $\hat{\theta} \to \theta$. In other words, $\hat{\theta}$ gets closer and closer to the true value θ as more data is collected. Again, we can test our Normal estimators for this property:

As $N \to \infty$, $\hat{\mu}_1$ = x_1 does not tend to μ .

As $N \to \infty$, $\hat{\mu}_2$ = $\frac{1}{2}(x_{(1)} + x_{(N)})$ does not tend to μ .

As $N \to \infty$, $\hat{\mu}_3$ = $\bar{x}$ does tend to μ .

As $N \to \infty$, $\hat{\mu}_4$ = $\frac{N+1}{N} \mu$ does tend to μ .

So the only consistent estimators are the third and fourth.

5.1.3 MINIMUM VARIANCE

$\hat{\theta}$ is a *minimum variance* estimator of θ if, for all possible estimators of θ from the same data set, it has the smallest variance. Consider our four possible estimators for μ:

$\text{Var}(\hat{\mu}_1)$ = $\text{Var}(x_1)$ = σ^2 , where σ^2 is the variance of X .

$\text{Var}(\hat{\mu}_2)$ is not easy to calculate, but it is clearly greater than σ^2 .

$\text{Var}(\hat{\mu}_3)$ = $\text{Var}(\bar{x})$ = σ^2/N .

$\text{Var}(\hat{\mu}_4)$ = $((N+1)/N)^2 \text{Var}(\bar{x})$ = $\sigma^2(N+1)^2/N^3$.

So $\hat{\mu}_3$ has the smallest variance of these four estimators. In fact, it has the smallest variance of any estimator of μ, and is the *minimum variance* estimator.

5.1.4 <u>LEAST SQUARED ERROR</u>

If we define the *mean squared error* $H = E((\hat{\theta}-\theta)^2)$, then $\hat{\theta}$ is the *least squared error* estimator of θ if H is a minimum for $\hat{\theta}$. Clearly, if $\hat{\theta}$ is both unbiased and minimum variance, then

$$H \;=\; E((\hat{\theta}-\theta)^2) \;=\; Var(\hat{\theta}),$$

which is a minimum.

So for our example problem of estimating the mean μ of a Normal random variable from a set of data values $x_1, x_2, \ldots x_N$, the least squared error estimator is the minimum variance unbiased estimator: $\hat{\mu} = \overline{x}$.

5.2 Estimating Variances

Suppose we wish to estimate the variance σ^2 of a Normal random variable from data $x_1, x_2, \ldots x_N$. What is the best estimator for this purpose ?

By definition, $\sigma^2 = E((X-\mu)^2)$, so it would clearly be sensible to consider the sums of the squares of the deviations of the data from the estimated mean $\overline{x}$:

$$S \;=\; \sum_{i=1}^{N} (x_i - \overline{x})^2 \; .$$

Taking the average square deviation from the mean gives us an estimator of σ^2:

$$\hat{\sigma}_1^2 \;=\; S/N \;=\; \frac{1}{N} \sum_{i=1}^{N} (x_i - \overline{x})^2 \; .$$

How does $\hat{\sigma}_1^2$ rate as an estimator of the true variance σ^2 - for example, is it unbiased ? To find its expected value $E(\hat{\sigma}_1^2)$, we need to use the following probability distribution:

If $x_1, x_2, \ldots x_N$ are all independently Normally distributed with variance σ^2, and

$$\chi^2 \;=\; S/\sigma^2 \;=\; \frac{1}{\sigma^2} \sum_{i=1}^{N} (x_i - \bar{x})^2 \;,$$

then this random variable χ^2 has a probability distribution with the same name ("chi-squared"), and with $N-1$ *degrees of freedom*. Its probability density function is shaped like:

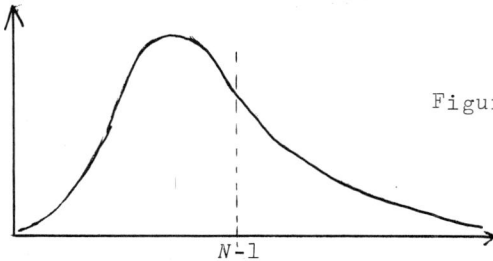

Figure 5.1

The mean $E(\chi^2) = N-1$. One way of confirming that this is reasonable is to consider the case when $N = 1$. Then $S = 0$ and $\chi^2 = 0$.

Now $\hat{\sigma}_1^2 \;=\; S/N \;=\; \sigma^2 \chi^2 / N$.

Therefore $E(\hat{\sigma}_1^2) \;=\; \frac{\sigma^2}{N} E(\chi^2) \;=\; \frac{N-1}{N} \sigma^2$.

This means that $\hat{\sigma}_1^2$ is not an unbiased estimator, since its expected value is not equal to σ^2. An alternative estimator is

$$s^2 \;=\; \frac{S}{N-1} \;=\; \frac{1}{N-1} \sum_{i=1}^{N} (x_i - \bar{x})^2 \;.$$

Clearly, $E(s^2) = \sigma^2$, and thus s^2 is an unbiased estimator. In general, s^2 is the estimator which will be used in preference to any other for estimating the variance of a Normal random variable.

5.3 Estimating Proportions

Suppose that we have N trials of a Binomial experiment with unknown success probability p, and that r of the trials turn out to be successes. An obvious estimator of p is

$$\hat{p} \quad = \quad r/N \quad = \quad \text{Proportion of successes achieved.}$$

$$E(\hat{p}) \quad = \quad E(r/N) \quad = \quad E(r)/N \quad = \quad Np/N \quad = \quad p \ .$$

So $\hat{p}$ is an unbiased estimator of p.

5.4 Confidence Intervals

Often it is not good enough to give just a single estimate $\hat{\theta}$ of an unknown parameter θ - we also would like some indication of the possible error in our estimate. To do this, we quote a *confidence interval* for θ, based on the given data.

For example, if we are estimating the mean μ of a Normal random variable based on N values, then a 95% confidence interval $[a,b]$ is such that $P(a \leq \mu \leq b) = 0.95$.

In general, a $C\%$ confidence interval is such that
$$P(\mu \text{ in the interval}) \quad = \quad C/100.$$
So we need to find the limits a and b of the interval - it would seem sensible to make them symmetrical about the best estimate of μ, $\bar{x}$:
$$a \quad = \quad \bar{x} - d, \quad b \quad = \quad \bar{x} + d \ .$$
Let $\alpha = 1 - C/100 = $ Probability that μ falls outside the confidence interval. Then we want:
$$P(|\bar{x} - \mu| > d) \quad = \quad \alpha \ .$$
Now, $\bar{x}$ is Normally distributed with mean μ and variance σ^2/N. Its probability density function is:

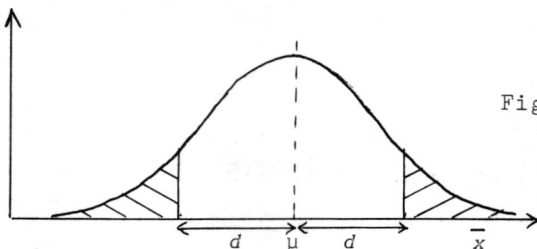

Figure 5.2

The probability that $\bar{x}$ lies more than d away from μ is given by the sum of the two shaded areas above, and this we want to be equal to α. To find the appropriate value of d, we must convert $\bar{x}$ to a Standard Normal variable Z by

$$Z = \frac{\bar{x} - \mu}{\sigma/\sqrt{N}} \ .$$

$$P(|\bar{x}-\mu| > d) \quad = \quad P(|Z| > d/(\sigma/\sqrt{N})) \quad = \quad \alpha \ .$$

Let z_α be such that $P(|Z| > z_\alpha) \quad = \quad \alpha$.

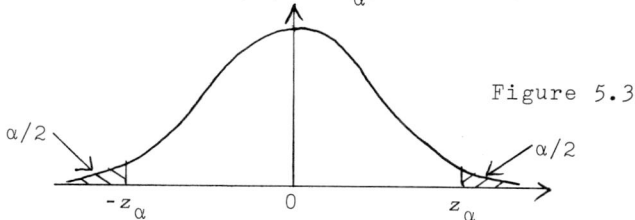

Figure 5.3

For example, if $\alpha = 0.05$, from Table 1 z_α is the value that gives a probability of $0.5 - 0.05/2 = 0.475$, which is 1.96.

So $d/(\sigma/\sqrt{N}) \quad = \quad z_\alpha$ means that $d \quad = \quad z_\alpha \sigma/\sqrt{N}$.

Therefore, the $1-\alpha$ confidence interval for μ is

$$\left[\bar{x} - z_\alpha \sigma/\sqrt{N} \ , \ \bar{x} + z_\alpha \sigma/\sqrt{N} \right] \ .$$

Example: Student IQ's are known to have a standard deviation of 20. 100 students are measured and found to have an average IQ of 105. Find a 95% confidence interval for the true mean student IQ.

$\alpha \quad = \quad 0.05$, so $z_\alpha \quad = \quad 1.96$ (from Table 1).

$\sigma \quad = \quad 20$ and $N \quad = \quad 100$.

$d \quad = \quad 1.96 \times 20/10 \quad = \quad 3.92$.

The 95% confidence interval is therefore $\left[101.08, \ 108.92 \right]$.

Had we wanted a 90% confidence interval, then we would have used $\alpha \quad = \quad 0.1$. From tables, z_α would be 1.645, so $d \quad = \quad 1.645 \times 20/10 \quad = \quad 3.29$, and the 90% confidence interval is $\left[101.71, \ 108.29 \right]$.

All the work we have done so far obviously needs us to know the value of the standard deviation σ exactly, but it is very often the case that we do not know σ any more than we know μ. Then we need to estimate σ^2 from the data, using the unbiased estimator s^2, but this introduces an extra source of uncertainty into the calculations. We can no longer use the

Standard Normal distribution to set up our confidence interval, and must use instead *Student's t-distribution*.

$$\text{If} \quad t \; = \; \frac{\overline{x} - \mu}{s/\sqrt{N}} \; ,$$

then t has a t-distribution with N-1 *degrees of freedom*. The degrees of freedom govern how close the t-distribution is to the Standard Normal:

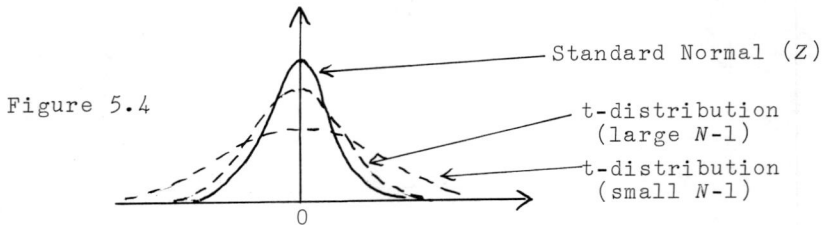

Figure 5.4

As N-1 increases, the t-distribution and the Standard Normal distribution become indistinguishable (i.e. s^2 becomes an exact estimator of σ^2).

We use the t-distribution to compute confidence intervals in exactly the same way as the Standard Normal. For a given error probability α, we need the "critical t-value", $t_{N-1}(\alpha)$, such that

$$P(|t| > t_{N-1}(\alpha)) \quad = \quad \alpha \; .$$

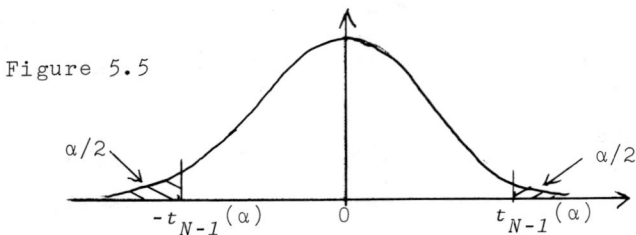

Figure 5.5

Example: Assuming student IQ's are Normally distributed with unknown mean and variance, find a 95% confidence interval for the mean IQ, given the following 6 measured IQ's:

$$102, \; 113, \; 98, \; 106, \; 92, \; 119 \; .$$

Estimated mean $\overline{x}$ = 105.0 .
Deviations from $\overline{x}$: -3.0, 8.0, -7.0, 1.0, -13.0, 14.0 .

Sum of squared deviations $S = 488.0$.
Estimated variance $s^2 = S/5 = 97.6$.
$$s = 9.88 .$$
Table 2 gives values of $t_{N-1}(\alpha)$ for different values of $N-1$ (degrees of freedom) and α (error probability). In our case $N-1 = 5$ and $\alpha = 0.05$, so that $t_{N-1}(\alpha) = 2.57$.

Confidence interval is $\overline{x} \pm t_{N-1}(\alpha)s/\sqrt{N}$
$$= 105.0 \pm 2.57 \times 9.88/\sqrt{6} = 105.0 \pm 10.365$$
$$= [94.635, 115.365] .$$

Another problem we may encounter is to find a confidence interval for the standard deviation σ. To do this we make use of the χ^2 distribution:
$$\chi^2 = S/\sigma^2 = (N-1)s^2/\sigma^2$$
has a χ^2 distribution with $N-1$ degrees of freedom.

So $\sigma^2 = (N-1)s^2/\chi^2$.
If χ^2 has a probability $1-\alpha$ of being between χ_1^2 and χ_2^2, then σ^2 has the same probability of being between
$$(N-1)s^2/\chi_2^2 \quad \text{and} \quad (N-1)s^2/\chi_1^2 .$$

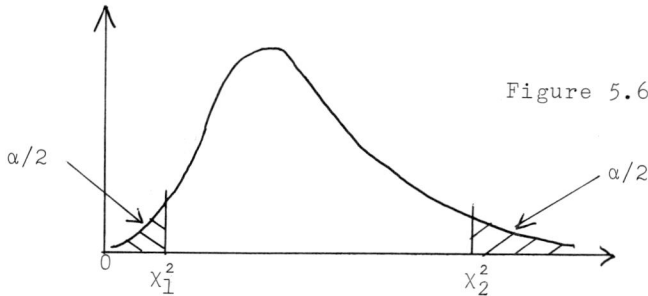

Figure 5.6

Table 3 gives values such that the probability of χ^2 being greater is equal to the given probability. For example, to get χ_2^2 we look up the table with probability $\frac{1}{2}\alpha$ and the correct degrees of freedom $(N-1)$. To get χ_1^2 we look it up with probability $1-\frac{1}{2}\alpha$.

Example: In the previous example, $s^2 = 97.6$ and there are 5 degrees of freedom. Find a 90% confidence interval for the true standard deviation.

From Table 3, for $\alpha = 0.1$, $\chi_1^2 = 1.145$ and $\chi_2^2 = 11.07$.
So a 90% confidence interval for σ^2 is
$$5 \times 97.6/11.07 \quad \text{to} \quad 5 \times 97.6/1.145$$
$$= [44.083, 426.20] .$$
A 90% confidence interval for the standard deviation σ is
$$[6.64, 20.64] .$$

Calculating confidence intervals for proportions requires, in theory, the use of the Binomial distribution. However, this can be tedious and it is often convenient to approximate the Binomial distribution by the Normal distribution, providing the number of trials N is sufficiently large.

For the Binomial distribution, the mean = Np and the variance = Npq.
$$\text{If } \hat{p} = r/N, \quad E(\hat{p}) = p$$
$$\text{and Var}(\hat{p}) = Npq/N^2 = pq/N .$$
Therefore, the estimated standard deviation of $\hat{p} = \sqrt{(\hat{p}\hat{q}/N)}$.

If we use a Normal approximation, the 1-α confidence interval for p will be given by
$$\hat{p} \pm z_\alpha \sqrt{(\hat{p}\hat{q}/N)} .$$

Example: An opinion poll samples 1000 voters, and finds that 400 say they will vote Conservative. What is a 95% confidence interval for the true proportion of Conservative voters in the whole population ?

$$N = 1000, \quad \hat{p} = 0.4, \quad \hat{q} = 0.6 .$$
Estimated standard deviation of $\hat{p} = \sqrt{(\hat{p}\hat{q}/N)} = 0.0155$.
So a 95% confidence interval is $0.4 \pm 1.96 \times 0.0155$
$$= [0.3696, 0.4304] .$$

Therefore, the opinion poll only tells us that the true proportion is quite likely to be somewhere between 37% and 43%. Note that to halve the width of the confidence interval, we should need to take four times the sample size, i.e. 4000 voters.

5.5 Estimating the Required Sample Size

Since the width of the confidence interval depends on the size N of the sample, it is possible to reverse the procedure, and estimate the size of sample required to produce a specified width for the confidence interval.

For example, if we know that student IQ's are Normally distributed with standard deviation 20, how large a sample is needed to estimate the mean IQ with an accuracy of ±1, with 95% confidence ?

We want the 95% confidence interval to be $\bar{x} \pm d$, with $d = 1$. Now, $d = z_\alpha \sigma/\sqrt{N}$, and rewriting this gives
$$N = z_\alpha^2 \sigma^2 / d^2 = 1.96^2 \times 20^2 / 1^2 = 1536.64 .$$

So a sample size of about 1537 would achieve the required accuracy. Of course, this calculation depended on us knowing the variance σ^2, and in general we may not know what this is. In many cases, however, we can obtain at least an approximate value from previous samples of similar data, or by practical considerations, or just by guesswork.

When estimating proportions, we know that the confidence interval is $\hat{p} \pm z_\alpha \sqrt{(\hat{p}\hat{q}/N)}$. For example, if we wish to conduct a poll to determine the percentage of Conservative voters with an accuracy of ±1% (with 95% confidence) then
$$d = z_\alpha \sqrt{(\hat{p}\hat{q}/N)} = 0.01 .$$
So $\qquad N = z_\alpha^2 \hat{p}\hat{q}/d^2 .$

Again, we do not know $\hat{p}$ beforehand, but we do know that the maximum value for $\hat{p}\hat{q}$ is when $\hat{p} = \hat{q} = \frac{1}{2}$. So a "conservative" value for N is
$$N = z_\alpha^2 \times \tfrac{1}{4}/d^2 = 1.96^2 \times \tfrac{1}{4}/0.01^2 = 9604 .$$
This is the required sample size for the desired accuracy. However, if we had some information about the probable value of $\hat{p}$, we could refine the estimate of N. For example, the Raving Loony Party usually polls about 1% of the vote, so that $\hat{p} \simeq 0.01$. So the required sample size for an accuracy of 1% is $\qquad N \simeq 1.96^2 \times 0.01 \times 0.99/0.01^2 = 380.3 .$

However, suppose we were asked to estimate the vote for the RLP with a 1% *relative* accuracy; i.e. with $d = 0.01 \times \hat{p}$, then

$$N = 1.96^2 \times 0.01 \times 0.99/(0.01 \times 0.01)^2$$
$$= 3,803,184 \ .$$

So it is important to be sure whether the desired accuracy is *absolute* (as a proportion of the total electorate) or *relative* (as a proportion of the party's vote).

5.6 Exercises

1. A random variable X is uniformly distributed from 0 to θ. A single value x is obtained from this distribution. Find an unbiased estimator for θ in terms of x.

2. A fisherman catches 6 fish, with the following weights in ounces: 12, 15, 11, 21, 17, 20.
   Estimate the mean and variance of the average weight of fish in this area, assuming they are Normally distributed.

3. The numbers of matches in match-boxes are approximately Normally distributed with standard deviation 5. Find a 90% confidence interval for the mean number of matches in a box based on the following sample:
        62, 67, 58, 63, 61, 59, 66, 58, 64 .
   Recompute the confidence interval assuming the standard deviation is unknown. Find a 90% confidence interval for the standard deviation.

4. A sample of 3600 voters produces 60 who vote for the Flat Earth Party. Find a 95% confidence interval for the proportion who will vote this way in the whole electorate. How large a sample would you need to take to get a relative accuracy of 5% ?

5. The number of jobs submitted to a Computer Centre each day is assumed to be Normally distributed. Over several days,

the following numbers have been recorded:

    126, 138, 104, 152, 161, 118, 111 .

Find a 95% confidence interval for the mean number of jobs per day. How much data would need to be collected to get an absolute accuracy of 5 jobs per day ? How much to get a relative accuracy of 2% ?

5.7 Computer Projects

1. Write a program to input a set of Normal data and produce confidence intervals for the mean and standard deviation. If it is not feasible to store the appropriate tables, the program should ask the user for the required critical points. The program should also be able to compute the required sample size for any desired accuracy in the mean.

2. Write a similar program (or adapt the first) to perform the same operations for proportions.

CHAPTER 6

Hypothesis Testing

6.1 Basic Principles

Statistics can be defined as the art (or science, if you prefer) of using numerical information to help in coming to conclusions about the real world, especially in uncertain situations. Note that statistics can never "prove" anything - all it can do is make some things seem rather likely or rather unlikely. To do this we use *hypotheses*. A hypothesis is a statement about reality, whose truth is unknown. We may wish to use statistical evidence either to accept a hypothesis (decide that it is likely to be true) or to reject it (decide it is unlikely to be true).

Here are three examples of the kinds of questions which we might want to answer using statistical methods:
1. Do these match-boxes contain at least 60 matches on average ?
2. Does Daz wash whiter than Brand X ?
3. Is this a fair coin ?

For every such question, we will in fact make up two opposing hypotheses, and try to use the statistical evidence to choose between them. One of these will be called the *null hypothesis* (its symbol is H_0), and will be the simpler of the two, usually involving some concept of equality. The other is called the *alternative hypothesis* (symbol H_1) and is usually more complex. Note that the question of which we should like to be true does not come into this. The selection of these two hypotheses will become clearer when we look at our examples again.

1. Let N be the average number of matches in a box. Define two hypotheses:

    Null hypothesis (H_0): "There are at least 60 matches per box on average",

    Alternative hypothesis (H_1): "There are less than 60 matches per box on average".

    Mathematically:
    $$H_0: \quad N \geq 60 ,$$
    $$H_1: \quad N < 60 .$$

    Note that H_0 includes the equality $N = 60$, which is why it is chosen as the null hypothesis.

2. Let D be the average whiteness produced by Daz, and X be that produced by Brand X.

    Null hypothesis (H_0): "Daz is the same as Brand X",

    Alternative hypothesis (H_1): "Daz washes whiter than Brand X".

    $$H_0: \quad D = X ,$$
    $$H_1: \quad D > X .$$

    Note that we are not even going to consider the possibility that $D < X$.

3. Let p be the probability of the coin landing heads.

    Null hypothesis (H_0): "This is a fair coin",

    Alternative hypothesis (H_1): "This is not a fair coin".

    $$H_0: \quad p = \tfrac{1}{2} ,$$
    $$H_1: \quad p \neq \tfrac{1}{2} .$$

In general, let us suppose there is some unknown parameter θ and we have a hypothetical value for it, say θ_0. Then our null hypothesis will look like:
$$H_0: \quad \theta = \theta_0 .$$
So in example 1, θ is N and θ_0 is 60. (Ignore the possibility that $N > 60$ in this case). In example 2, θ will be the difference $D-X$, which makes θ_0 equal to zero. And in example 3, θ is p and θ_0 is $\tfrac{1}{2}$.

The form of the alternative hypothesis depends on the

particular problem. Commonly, it will be one of the following:

$$\text{a)} \quad H_1: \; \theta \; < \; \theta_0$$
$$\text{b)} \quad H_1: \; \theta \; > \; \theta_0$$
$$\text{c)} \quad H_1: \; \theta \; \neq \; \theta_0 \;.$$

Case c) is a combination of a) and b) - later on we shall
talk about "two-sided" tests when H_1 is of this form.

In principle at least, all this work of choosing the
hypotheses to be tested is done before any statistical
evidence is gathered. Then we need to decide how we are going
to let the data tell us whether to accept H_0 and reject H_1, or
accept H_1 and reject H_0. If we think of ourselves as playing
a kind of game against nature, there are four possible results,
as follows:

<u>We decide</u>

Reality	Accept H_0	Accept H_1
H_0 true	✓	×
H_1 true	×	✓

So there are two different ways of getting the wrong
result in hypothesis testing - these two kinds of error are
given special names.

A *Type I Error* is when we accept H_1 (and reject H_0) when H_0 is
in fact true.

A *Type II Error* is when we accept H_0 (and reject H_1) when H_1
is in fact true.

Now of course we will not know whether or not we have
committed one of these errors, because we do not know the
true state of reality. What we can do, making various
assumptions, is to compute the probabilities of making these
errors, conditional on the state of reality. These
probabilities are usually called α and β.

For a particular testing set-up, define α to be the
probability of a Type I error, and β to be the probability of
a Type II error.

$$\alpha = P(\text{Accept } H_1 | H_0 \text{ is true}),$$
$$\beta = P(\text{Accept } H_0 | H_1 \text{ is true}).$$

Ideally, we should like to make both α and β as small as possible. Unfortunately, if we arrange to decrease α we will increase β, and vice versa. We are therefore compelled to strike a balance between these two types of error, and there is no overall "best" answer to the question of how to strike this balance.

Normally, our hypothesis testing is based upon selecting a given maximum value of α. It is usually easier to work with α rather than β, because α is computed assuming H_0 is true, and H_0 is a simpler hypothesis than H_1. To show how this works, let us consider a simple example.

Example: Suppose we have a penny, which we suspect may be biased to give less than the correct number of heads on average. The parameter we are interested in is p, the probability of the coin landing heads. Our hypotheses are:
$$H_0: p = \tfrac{1}{2} \quad \text{vs.} \quad H_1: p < \tfrac{1}{2}.$$
To find out which hypothesis to accept, we will toss the coin 10 times and count the number of heads we get - call this r. Beforehand, we will choose an integer d, and we will

Accept H_0 if $r > d$ (coin is unbiased), or

Accept H_1 if $r \leq d$ (coin is biased).

The problem thus boils down to choosing a suitable value of d. Let us investigate the Type I error probabilities for various values of d.

Now $P(\text{Type I error}) = P(\text{Accept } H_1 | H_0 \text{ true})$
$$= P(r \leq d | p = \tfrac{1}{2}).$$

We may obtain these probabilities from the Binomial probability distribution, with $n = 10$ and $p = \tfrac{1}{2}$.

Number of heads (r)	$P(r)$	$F(r) = P(\leq r)$
0	0.0010	0.0010
1	0.0098	0.0108
2	0.0439	0.0547
3	0.1172	0.1719
4	0.2051	0.3770
5	0.2461	0.6231

Suppose we choose $d = 3$, i.e. accept H_0 if we get more than 3 heads and H_1 otherwise.

$$P(\text{Type I error}) = P(\text{Accept } H_1 | H_0 \text{ true})$$
$$= P(r \leq 3 | p = \tfrac{1}{2}) = F(3) = \underline{0.1719} .$$

Thus with $d = 3$ there is a 17% chance of wrongly deciding the coin is biased when it is really quite fair. In general, for any choice of d,

$$P(\text{Type I error}) = P(r \leq d | p = \tfrac{1}{2}) = F(d) .$$

The normal procedure for choosing d is to decide on an acceptable level for the Type I error probability, so that it is $\leq \alpha$ say. If we set $\alpha = 0.1$, for example, then we must choose d to give $F(d) \leq 0.1$, which from the above table means d should be 2. Therefore we accept H_0 if we get more than 2 heads and accept H_1 if we get 2 or fewer. What we are saying is "If p really is $\tfrac{1}{2}$, then there is no more than a 10% probability of getting 2 or fewer heads from 10 tosses. So if that happens, we shall assume that p is in fact less than $\tfrac{1}{2}$".

The value α which controls the P(Type I error) is called the *level of significance* for the test.

If we set up this test, accepting H_0 if $r > 2$ and H_1 if $r \leq 2$, what is the P(Type II error) ? This will depend on the true value of p, the unknown probability of heads. Suppose p were equal to $1/3$ - then we can tabulate the Binomial probabilities for the number of heads obtained using $n = 10$ and $p = 1/3$:

Number of heads (r)	$P(r)$	$F(r)$
0	0.0173	0.0173
1	0.0867	0.1040
2	0.1951	0.2991
3	0.2601	0.5592
4	0.2276	0.7868

So $P(\text{Type II error}) = P(\text{Accept } H_0 | H_1 \text{ true})$
$$= P(r > 2 | p = 1/3) = 1 - F(2)$$
$$= \underline{0.7009} .$$

This means that if the probability of heads (p) is really $1/3$, there is a 70% chance that our test will lead us to accept (wrongly) the null hypothesis that p is $\tfrac{1}{2}$. Normally we talk about the *power* of a test for a particular value of the parameter.

Power $= P(\text{Accept } H_1 | H_1 \text{ true}) = 1 - P(\text{Type II error})$. This is the probability of correctly deciding that the null hypothesis is false, i.e. of discriminating between the null value and the true value of the parameter. In this case, with $p = 1/3$, the power $= 0.2991$.

The power of the test varies with the exact value of p. We may plot a *power function*, which leads to a graph of power as a function of p, looking something like:

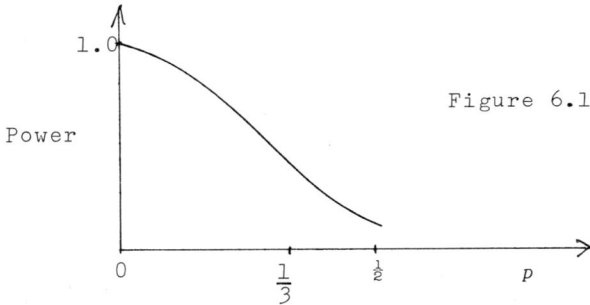

Figure 6.1

Clearly, the further p is from the value $\frac{1}{2}$, the higher the probability of correctly deciding that p is not equal to $\frac{1}{2}$ with this test. As p approaches $\frac{1}{2}$ our test is clearly not very "powerful" - i.e. it does not have a very good chance of telling us that p is not $\frac{1}{2}$. Ideally, our perfect test would have a power function shaped like this:

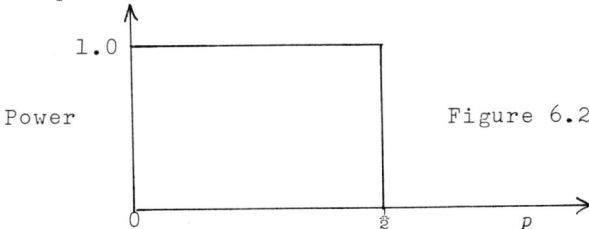

Figure 6.2

At this point, it is probably worth giving a glossary of some of the main jargon terms used in hypothesis testing.

1. Level of significance

The specified probability α which the $P(\text{Type I error})$ must not exceed. We set up the test so that $P(\text{Type I error}) \leq \alpha$.

2. Power

   For a given value of the unknown parameter, the power of the test is $P(\text{Accept } H_1 | H_1 \text{ true})$.

3. Statistic

   A statistic is a single numerical value, computed from the results of an experiment, which is used to decide between H_0 and H_1. (For example, in the coin-tossing experiment the statistic used was r, the number of heads).

4. Critical Region

   The critical region is the set of values of the statistic for which we accept H_1. (In the coin-tossing experiment, this is $r \leq 2$, i.e. $\{0,1,2\}$). The set of values for which we accept H_0 is the "acceptance region". (In our case, this is $r > 2$, i.e. $\{3,4,5,6,7,8,9,10\}$).

6.2 General Strategy for Hypothesis Testing

   In all kinds of hypothesis testing cases, however complex, the following general strategy should be carried out:

1. Decide on the hypotheses to be tested, H_0 and H_1.

2. Assume H_0 is true; then calculate the probability distribution for the statistic we are going to use to test the hypotheses.

3. For a given level of significance α, choose the critical region for the test statistic, so that $P(\text{Type I error}) \leq \alpha$.

4. Perform the experiment and record the results.

5. Compute the value of the test statistic from the results of the experiment.

6. Compare the value of the test statistic with the critical region and decide whether to accept H_0 or H_1.

   Notice that in this scheme the experiment is only

performed as step 4 - all the work of setting up the test is done first. In practice, of course, especially when doing exercise problems, the experimental results may already be available before setting up the test. However, it is important to ignore them until the right point in the general hypothesis testing scheme.

Example: A random number generator is supposed to produce a random number X uniformly distributed between 0 and 1. However, it is suspected that in fact the upper limit may be less than 1.0. A number has been generated and turns out to be 0.03178. How do we test whether or not the upper limit is truly 1.0, with a significance level of 0.05 ?

Step 1: The parameter we are interested in is the upper limit to the uniform distribution, θ. The two hypotheses are:

$$H_0: \theta = 1.0 \text{ vs. } H_1: \theta < 1.0 .$$

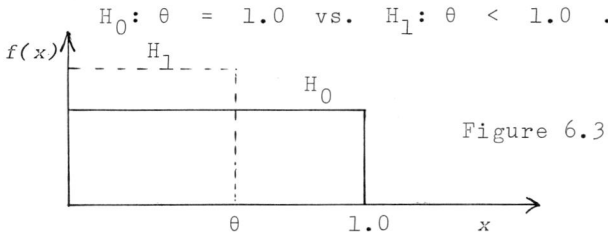

Figure 6.3

Step 2: Assume H_0 is true - i.e. $\theta = 1.0$. The test statistic is going to be just X, the number obtained from the random number generator. To carry out the test, we want a critical value d, so that we

Accept H_0 if $X \geq d$, and
Accept H_1 if $X < d$.

So the critical region is $[0, d)$. The probability distribution of X if H_0 is true is uniform:

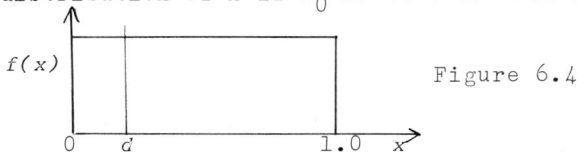

Figure 6.4

Step 3: The significance level $\alpha = 0.05$. We want P(Type I error) $= \alpha$. In the Binomial case we could not make this probability exactly equal to α, but now we can, because we are dealing with a continuous random variable. So we want $P(X < d) = 0.05$. For this uniform distribution, with $\theta = 1.0$, $P(X < d) = d$ (see the shaded area below).

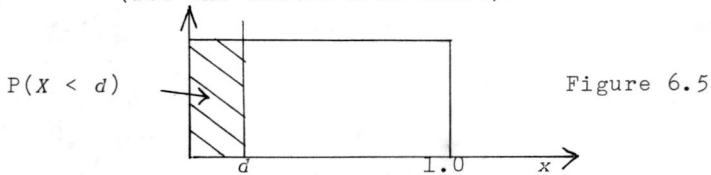

P($X < d$)

Figure 6.5

So we just need $d = 0.05$. The critical region is $[0, 0.05)$ and the acceptance region $[0.05, 1.0]$.

Step 4: Perform experiment - already done.

Step 5: Calculate test statistic - $X = 0.03178$.

Step 6: Comparing the value of X with the critical region, we accept H_1, that $\theta < 1.0$. In other words, there is a less than 5% chance that this value of X would have occurred if θ were equal to 1.0 .

To plot the power function for this test, we just need $P(X < 0.05 | \theta)$ for all possible θ.

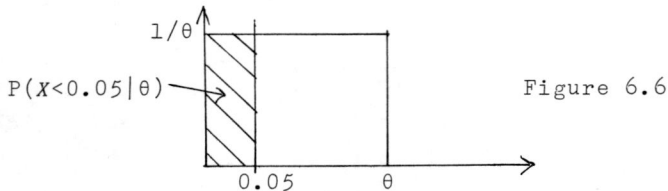

$1/\theta$

P($X < 0.05 | \theta$)

0.05 θ

Figure 6.6

So $P(X < 0.05 | \theta) = 0.05/\theta$.
Hence we may plot the power function:

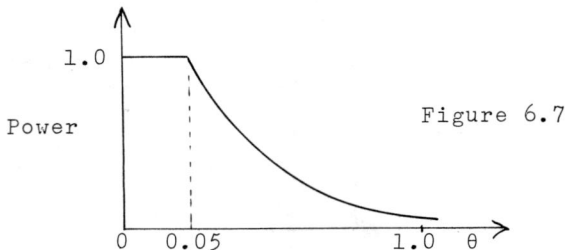

1.0

Power

Figure 6.7

0   0.05   1.0 θ

So if the true value of θ were 0.5, for example, then the probability of correctly accepting H_1 would be $0.05/0.5 = 0.1$ or 10%.

6.3 Hypothesis Testing with Normal Distribution

Hypothesis testing applications where the underlying probability distribution is assumed to be Normal are very common. Of course, the Normal distribution has two parameters - μ (the mean) and σ (the standard deviation). To simplify matters, for the present we shall assume that we know σ exactly, and are just unsure of the value of μ.

Example: A machine makes ball-bearings with a nominal diameter of 10 mm and a known standard deviation of 2 mm. The machine may be suffering from a malfunction, in which case the mean diameter will be less than 10 mm. One ball-bearing is selected at random and its diameter measured. How small must it be for us to decide that the machine is malfunctioning, with a level of significance equal to 0.05 ?

Step 1: Our two hypotheses are:
$$H_0: \quad \mu = 10 \quad \text{vs.} \quad H_1: \quad \mu < 10 .$$

Step 2: Assume H_0 is true: $\mu = 10$. The test statistic is just X, the diameter of the randomly-chosen ball-bearing. This will be Normally distributed with mean 10 and standard deviation 2, if H_0 is true.

Step 3: Choose a "critical diameter" d so that we
Accept H_0 if $X \geq d$, and
Accept H_1 if $X < d$.
The critical region is thus $(0, d)$. Choose d so that
P(Type I error) = 0.05, i.e.
$$P(X < d \mid \mu = 10) = 0.05 .$$
We can see what this means graphically:

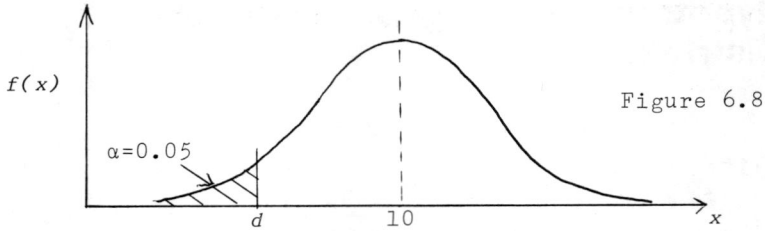

Figure 6.8

We choose d so that the shaded area to the left of d is equal to 0.05. Converting to the Standard Normal:

Let $z_d = (d-\mu)/\sigma = (d-10)/2$.

From Standard Normal tables, $z_d = -1.645$.

So $d = 2\times(-1.645) + 10 = 6.71$.

The critical region for the test is $(0, 6.71)$.

<u>Steps 4&5</u>: Measure the ball-bearing, and find $X = 8.3$.

<u>Step 6</u>: The value 8.3 is outside the critical region, so we accept H_0. In other words, there is not enough evidence for us to conclude that the machine is malfunctioning.

To plot the power function in this case, we need to compute, for each possible value of μ, Power = $P(\text{Accept } H_1 | \mu)$ = $P(X < d | \mu)$. Let us tabulate this for a few values of μ:

True mean μ	$z_d = \dfrac{6.71-\mu}{2}$	$P(Z < z_d)$ = Power
2.0	2.355	0.9907
4.0	1.355	0.9123
6.0	0.355	0.6387
8.0	-0.645	0.2595

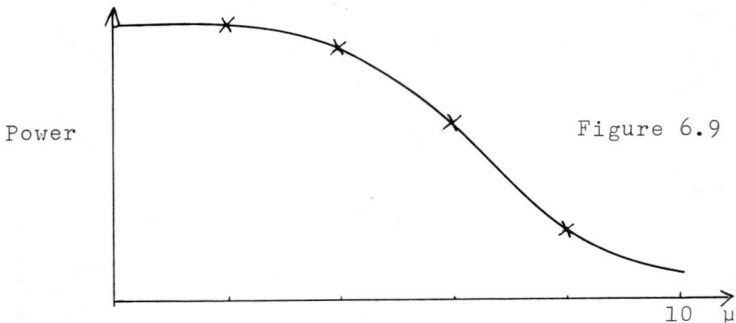

Figure 6.9

Power

10 μ

104

The above test is called a "one-sided" test because we have specifically excluded the possibility that the mean may actually be greater than 10. Suppose, however, that the machine may malfunction by giving a mean diameter greater than 10 as well as by giving one less than 10. In this case we need to set up a "two-sided" test.

$$H_0: \mu = 10 \quad \text{vs.} \quad H_1: \mu \neq 10 \ (\mu < 10 \text{ or } \mu > 10).$$

To specify the critical region we need two critical diameters d_1 and d_2, and the decision will be:

$$\text{Accept } H_0 \text{ if } d_1 \leq X \leq d_2 \text{, and}$$
$$\text{Accept } H_1 \text{ if } X < d_1 \text{ or } X > d_2.$$

Clearly it is sensible to arrange d_1 and d_2 symmetrically about 10. If we use a total level of significance of 0.05, we need to split this between the two sides of the distribution.

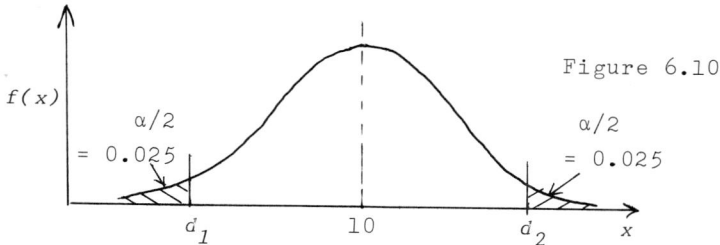

Figure 6.10

To find d_1, let $z_1 = (d_1-10)/2$.
We want $P(Z < z_1) = 0.025$, which from tables means $z_1 = -1.96$, i.e. $d_1 = 6.08$.
Similarly, $z_2 = (d_2-10)/2 = 1.96$, and $d_2 = 13.92$.
The acceptance region is thus $[6.08, 13.92]$. The power of this two-sided test for any value of μ is

$$P(X < d_1) + P(X > d_2) = P(Z < \frac{6.08-\mu}{2})$$
$$+ P(Z > \frac{13.92-\mu}{2})$$

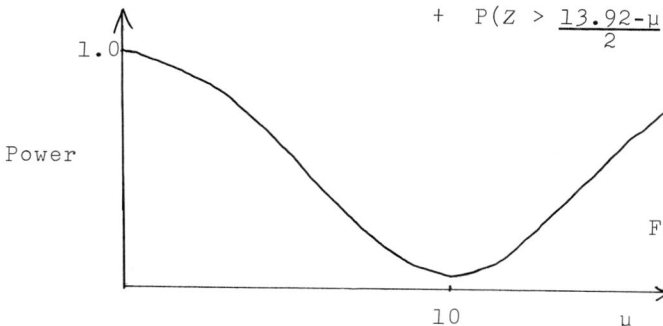

Figure 6.11

Obviously, this test is not very powerful. It does not have much chance of detecting small differences from 10 mm in the mean diameter. To improve the power of the test, it is sensible to extend our experiment to include the diameters of more than one ball-bearing. Suppose we measure 4 ball-bearings.

The test statistic will be $\overline{x}$, the average of the 4 diameters measured. We know the variance of $\overline{x}$ will be $4.0/4 = 1.0$, and the standard deviation is 1.0. Again, we have a two-sided test with critical values d_1 and d_2, but with a smaller standard deviation for the test statistic.

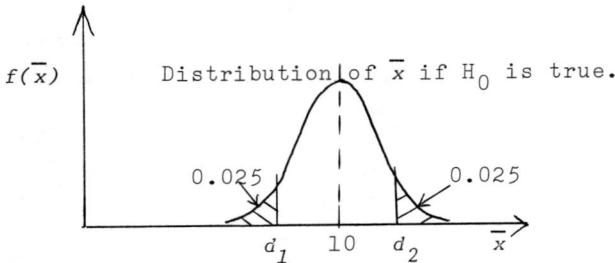

Figure 6.12

Choose $z_1 = (d_1 - 10)/1$ so that $P(Z < z_1) = 0.025$. Again $z_1 = -1.96$, but this time $d_1 = 8.04$. In the same way $d_2 = 11.96$, and the acceptance region for $\overline{x}$ is $[8.04, 11.96]$.

The power of the test is higher than before, because d_1 and d_2 are closer to 10.

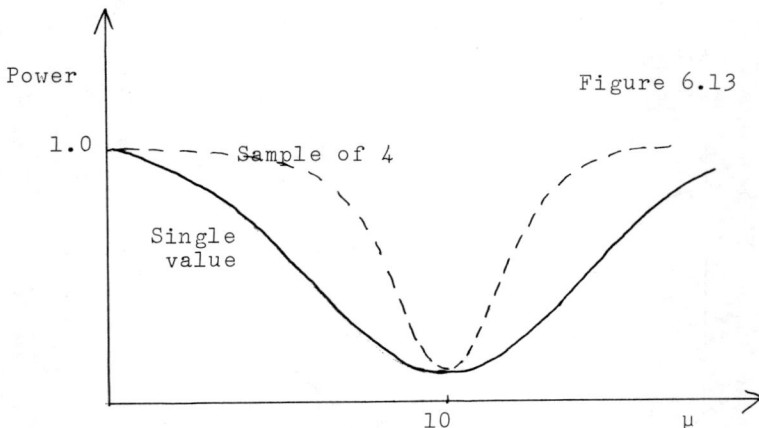

Figure 6.13

106

6.4 Exercises

1. A "psychic" medium is being tested for clairvoyance. A
    large pack of cards consists of only 5 different symbols,
    and the medium has to guess the symbol on a card which is
    dealt face down in front of her. The experiment is
    repeated 5 times. How many cards must she get right for
    us to decide she is clairvoyant, with a 5% significance
    level ?

2. "Yummo" catfood is being compared with two other brands.
    6 random cats are each exposed to 3 identical dishes of
    catfood, and the brand which is eaten first is recorded
    for each cat. How many cats must eat "Yummo" first for
    us to conclude that it is preferred to the others, with a
    10% significance level ?

3. Projector lamps are known to have lifetimes which are
    negatively exponentially distributed. Their mean lifetime
    is supposed to be 50 hours. Devise a test to see if the
    actual mean is less than this, using a significance level
    of 5%. If a projector lamp blows after 10 hours, would
    you conclude that the mean lifetime was less than 50 hours?

4. The number of matches in a box is (approximately) Normally
    distributed with standard deviation 8 matches. A box is
    supposed to hold 60 matches on average. A man finds his
    box has only 40 matches in it - is he justified in
    concluding, with a 1% significance level, that the mean is
    really less than 60 ?

5. The mean petrol consumption of new cars from an assembly
    line is meant to be 35 m.p.g., with a standard deviation
    of 2.5 m.p.g. Assuming Normality, devise a test using a

sample of 10 cars to detect differences from the assumed
mean with a significance level of 1%. 10 cars are found
to have values of 36.2, 34.1, 33.3, 32.5, 33.1, 34.3, 35.0,
33.7, 32.9 and 34.4 m.p.g. What conclusion would you
draw ? Sketch the power function for values of the mean
between 30 and 40 m.p.g.

6. Two production lines produce ball-bearings of nominal
   diameter 30 mm, with a variance of 10 mm^2. Assuming
   Normality and a significance level of 5%, detect any
   difference between the two lines based on the following
   data:
   Line A: 35, 33, 36, 30, 37
   Line B: 32, 34, 31, 35, 33 .
   Is there a difference between the nominal diameter and
   the actual mean ?

6.5 Computer Projects

1. Write a program to give the critical region for a one-
   sided Binomial hypothesis testing problem, given the
   number of trials, hypothetical success probability and
   significance level. It should test either
   a) $H_0: p = p_0$ vs. $H_1: p < p_0$
   or b) $H_0: p = p_0$ vs. $H_1: p > p_0$.

2. Write a program which will carry out two-sided Normal
   tests, given the number in the sample, the hypothetical
   mean, known standard deviation, critical value of the
   Standard Normal distribution for the desired significance
   level, and the experimental data.

CHAPTER 7

T Tests

In Chapter 5 we saw how Student's t-distribution was used to produce confidence intervals for Normal random variables when the variance was estimated from the data. In hypothesis testing, we also use the t-distribution when testing Normal random variables whose variance is unknown. There are various ways in which we can use the t-distribution in hypothesis testing, depending on the problem.

7.1 Simple Hypotheses about the Mean

Suppose we take a sample of N values $x_1, x_2, \ldots x_N$ from a random variable X which is Normally distributed with mean μ and variance σ^2, and that the null hypothesis is

$$H_0: \mu = \mu_0 .$$

We can consider three possible alternative hypotheses:

1. $H_1: \mu < \mu_0$
2. $H_1: \mu > \mu_0$
3. $H_1: \mu \neq \mu_0 .$

The first two lead to one-sided tests, while the third gives a two-sided test and is in fact a combination of the other two. The test we carry out will thus depend mainly on the nature of the alternative hypothesis.

$$H_0: \mu = \mu_0 \text{ vs. } H_1: \mu < \mu_0 .$$

Calculate $\bar{x} = \dfrac{1}{N} \sum_{i=1}^{N} x_i$

and $s^2 = \dfrac{1}{N-1} \sum_{i=1}^{N} (x_i - \bar{x})^2$

and hence find $t = \dfrac{\bar{x} - \mu_0}{s/\sqrt{N}} .$

If we assume H_0 is true ($\mu = \mu_0$), then t has a t-distribution with $N-1$ degrees of freedom. We specify a level of significance for the test:

$$P(\text{Accept } H_1 | H_0 \text{ true}) = \alpha .$$

The statistic on which we shall base the test will be the value of t, as calculated above. Clearly, large positive values of t support H_0 and large negative values support H_1. We need a critical value, T say, so that we accept H_0 if $t \geq T$ and H_1 if $t < T$. We want

$$P(t < T | H_0 \text{ true}) = \alpha .$$

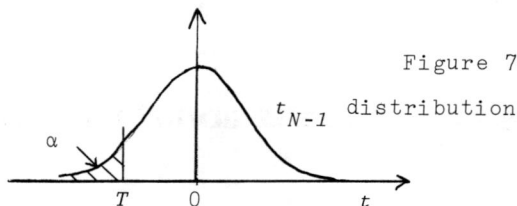

Figure 7.1

t_{N-1} distribution

Given α and the number of degrees of freedom ($N-1$), we can obtain T from tables. Table 2 gives critical values of the t-distribution, but it is based on a two-sided test, in which

$$P(|t| > T | H_0 \text{ true}) = \alpha .$$

To get critical values for a one-sided test from Table 2, we just pretend we are doing a two-sided test with significance level 2α, and discard the upper half of the test. Thus we will get a critical value which is actually the negative of the value from the table.

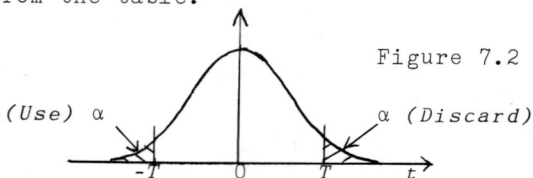

Figure 7.2

(Use) α α (Discard)

Example: Match-boxes are supposed to contain, on average, 60 matches per box. A customer complains that this is not in fact true. To check this, 10 boxes are opened at random and the number of matches in each box counted. The results are:

$$55, \ 62, \ 53, \ 51, \ 63, \ 49, \ 54, \ 52, \ 57, \ 55 \ .$$

Is the customer right ?

$$H_0: \mu = 60 \quad \text{vs.} \quad H_1: \mu < 60 \ .$$

$$\bar{x} = 551/10 = 55.1 \ .$$

$$s^2 = \frac{1}{9} \sum_{i=1}^{10} (x_i - 55.1)^2 = 20.322 \ . \quad s = 4.508 \ .$$

$$t = \frac{\bar{x} - 60}{s/\sqrt{10}} = -3.437 \ .$$

If we choose $\alpha = 0.05$, then we need to look up Table 2 with a 10% error probability (because of only doing a one-sided test) and 9 degrees of freedom. We find a value of 1.83, and because of the form of H_1 we in fact want $T = -1.83$.

i.e. Accept H_0 if $t \geq -1.83$,
Accept H_1 if $t < -1.83$.

The value of t we calculated from the data was -3.437, which is less than -1.83, so we accept H_1. This means that we believe the customer is right, and the mean number of matches per box is less than 60.

7.1.2 $\qquad H_0: \mu = \mu_0 \quad \text{vs.} \quad H_1: \mu > \mu_0 \ .$

This is a one-sided test, as in the previous case, except that we $\qquad$ Accept H_0 if $t \leq T$,
Accept H_1 if $t > T$.

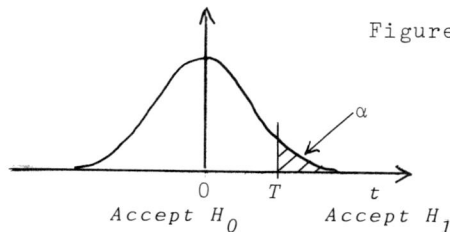

Figure 7.3

111

$H_0: \mu = \mu_0$ vs. $H_1: \mu \neq \mu_0$.

This is a two-sided test, a combination of the previous two
cases. The values of T are given directly by Table 2.

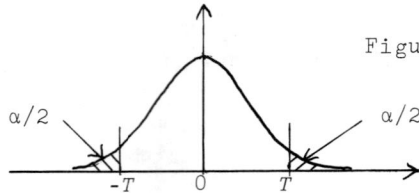

Figure 7.4

Example: A machine is supposed to turn out components with a
mean length of 50 mm. 6 components are measured,
and found to have lengths of 43, 46, 49, 51, 42 and
44 mm. Is the machine defective ?

$$H_0: \mu = 50 \quad \text{vs.} \quad H_1: \mu \neq 50 .$$

$\overline{x}$ = 45.833 .
s^2 = 12.567 ; s = 3.545 .
t = -2.879 .

With α = 0.05, from tables we get T = 2.57 with 5 degrees of
freedom. Since $|t| > 2.57$, we must accept H_1, that $\mu \neq 50$.
So the machine is defective.

The program that follows performs a two-sided t test,
and stores the table of critical t-values.

```
program ttest(input,output,tfile);
{ Program to perform 2-sided t test }
const maxdata = 30;
      maxsig = 6;
      maxdf = 33;

type data = array[1..maxdata] of real;
     posint = 0..maxint;
     tabletype = array[1..maxdf,1..maxsig] of real;
     dftype = array[1..maxdf] of posint;
     sigtype = array[1..maxsig] of real;

var ttable : tabletype;
    x : data;
    dftable : dftype;
    sigtable : sigtype;
    i,n : posint;
    alpha,xbar,s,t,tcrit,mu : real;
    tfile : text;
```

```
procedure settable;
{ Procedure to read in t table values from file. }
var i,j : posint;
begin
  reset(tfile);
  for j := 1 to maxsig do read(tfile,sigtable[ j ]);
  for i := 1 to maxdf do
  begin
    read(tfile, dftable[ i ]);
    for j := 1 to maxsig do read(tfile,ttable[ i,j ]);
  end;
end;

function tablet(df : posint; alpha : real) : real;
{ Function to compute critical t value from table. }
var i,j : posint;
    t1,t2,fact : real;
begin
  i := 2; j := 2;
  while (dftable[ i ]<df) and (i<maxdf) do i := i + 1;
  i := i - 1;
  while (sigtable[ j ]>alpha) and (j<maxsig) do j := j + 1;
  j := j - 1;
  fact := (alpha-sigtable[ j ])/(sigtable[ j+1 ]
          - sigtable[ j ]);
  t1 := ttable[ i,j ]+ fact*(ttable[ i,j+1 ]-ttable[ i,j ]);
  t2 := ttable[ i+1,j ] + fact*(ttable[ i+1,j+1 ]
          -ttable[ i+1,j ]);
  tablet := t1 + (df-dftable[ i ])*(t2-t1)/(dftable[ i+1 ]
          -dftable[ i ]);
end;

function tcalc(var x : data; mu : real; n : posint) : real;
{ Function to compute t statistic from the data. }
var t : real;
    i : posint;
begin
  xbar := 0.0; s := 0.0;
  for i := 1 to n do xbar := xbar + x[ i ];
  xbar : = xbar/n;
  for i := 1 to n do s := s + sqr(x[ i ]-xbar);
  s := s/(n-1); s := sqrt(s);
  t := (xbar-mu)/(s/sqrt(n));
  writeln; writeln('Mean of data = ',xbar:10:4);
  writeln('Standard deviation = ',s:10:4);
  writeln('t value = ',t:10:4);
  tcalc := t;
end;

begin
  settable;
  writeln; writeln('Input number of data values: ');
  read(n); writeln;
  writeln('Input data'); writeln;
  for i := 1 to n do read(x[ i ]);
  writeln; writeln('Input significance level : ');
  read(alpha); writeln;
```

```
writeln('Input hypothetical mean : ');
read(mu); writeln;
t := tcalc(x,mu,n);
tcrit := tablet(n-1,alpha);
writeln('Critical t value is ',tcrit:10:4);
if abs(t) < tcrit
   then writeln('Accept Null Hypothesis')
   else writeln('Accept Alternative Hypothesis');
end.
```

7.2 Differences between Sample Means

The t-distribution can be used to test whether or not
the means of two random variables are the same. In fact,
this can be done using two different approaches.

7.2.1 <u>THE TWO-SAMPLE T-TEST</u>

Suppose we have two random variables, both Normally
distributed: X_1 with mean μ_1 and X_2 with mean μ_2, and both
with the same variance σ^2. Suppose also that we do not know
μ_1, μ_2 or σ^2. We wish to test

$$H_0: \mu_1 = \mu_2 \quad \text{vs.} \quad H_1: \mu_1 \neq \mu_2$$

or, $\qquad H_0: \mu_1 - \mu_2 = 0 \quad \text{vs.} \quad H_1: \mu_1 - \mu_2 \neq 0 .$

We take a sample of n_1 values $x_{11}, x_{12}, \ldots x_{1n_1}$ from X_1 ,

and a sample of n_2 values $x_{21}, x_{22}, \ldots x_{2n_2}$ from X_2 .

Calculate: $\quad \overline{x}_1 = \dfrac{1}{n_1} \sum_{i=1}^{n_1} x_{1i}$, $\quad s_1^2 = \dfrac{1}{n_1-1} \sum_{i=1}^{n_1} (x_{1i} - \overline{x}_1)^2$

$$\overline{x}_2 = \dfrac{1}{n_2} \sum_{i=1}^{n_2} x_{2i} , \quad s_2^2 = \dfrac{1}{n_2-1} \sum_{i=1}^{n_2} (x_{21} - \overline{x}_2)^2 .$$

To set up a t statistic we need to estimate the
variance σ^2. We do this by means of the *pooled variance*,
which is an average of the two estimated variances s_1^2 and s_2^2,
weighted by their degrees of freedom.

$$s^2 = \frac{(n_1-1)s_1^2 + (n_2-1)s_2^2}{n_1 + n_2 - 2} \quad .$$

The divisor is n_1+n_2-2 because 2 degrees of freedom have been "used up" in estimating $\bar{x}_1$ and $\bar{x}_2$. We now have n_1+n_2-2 degrees of freedom left for the estimation of the variance.

Our test is going to be based on the difference between the two sample means, $\bar{x}_1-\bar{x}_2$. To form a t statistic, we need to estimate the variance of this difference.

$$\begin{aligned}
\text{Var}(\bar{x}_1-\bar{x}_2) &= \text{Var}(\bar{x}_1) + \text{Var}(\bar{x}_2) \\
&= \frac{\sigma^2}{n_1} + \frac{\sigma^2}{n_2} = \sigma^2\left(\frac{1}{n_1} + \frac{1}{n_2}\right) .
\end{aligned}$$

The standard deviation of $\bar{x}_1-\bar{x}_2 = \sqrt{\left(\sigma^2\left(\frac{1}{n_1} + \frac{1}{n_2}\right)\right)}$.

Therefore, if H_0 is true, the statistic

$$t = \frac{\bar{x}_1 - \bar{x}_2}{\sqrt{\left(s^2\left(\frac{1}{n_1} + \frac{1}{n_2}\right)\right)}}$$

has a t-distribution with n_1+n_2-2 degrees of freedom.

But we are making a very important assumption - namely that X_1 and X_2 both have the same variance σ^2. We need to carry out a preliminary test to validate this assumption before continuing with the main t-test. Assuming X_1 has variance σ_1^2 and X_2 has variance σ_2^2, we test

$$H_0: \sigma_1^2 = \sigma_2^2 \quad \text{vs.} \quad H_1: \sigma_1^2 \neq \sigma_2^2 \quad .$$

Clearly we use the sample variances s_1^2 and s_2^2 to test this; in fact, the test is based on the *variance ratio*

$$f = s_1^2/s_2^2 \quad .$$

This ratio is always arranged so that the largest variance is on top, so that $f \geq 1$. If H_0 is true, then f has a probability distribution called an f-distribution, looking like:

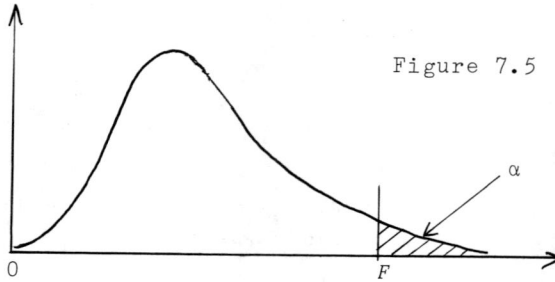

Figure 7.5

The shape of this distribution depends on two values for degrees of freedom: n_1-1 and n_2-1 for the two variances.

Table 4 gives values of $F_{n_2-1}^{n_1-1}(\alpha)$, the critical value such that

$$P(f > F_{n_2-1}^{n_1-1}(\alpha)) \quad = \quad \alpha \quad \text{if } H_0 \text{ is true.}$$

If we accept H_0, that the two variances are equal, then we may proceed to the main t-test.

Example: Two production lines fill cans with peas. Do they both fill with the same number, on average ?

Line 1: 159, 167, 173, 148, 162
Line 2: 183, 179, 174, 186, 189, 175 .

Line 1

$$\bar{x}_1 \;=\; 161.8 \qquad \sum_i (x_{1i} - \bar{x}_1)^2 \;=\; 350.8 .$$
$$s_1^2 \;=\; 87.7 .$$

Line 2

$$\bar{x}_2 \;=\; 181.0 \qquad \sum_i (x_{2i} - \bar{x}_2)^2 \;=\; 182.0 .$$
$$s_2^2 \;=\; 36.4 .$$

To test equality of variances, form $f = 87.7/36.4 = 2.41$. Degrees of freedom are 4 and 5 for the top and bottom. From Table 4, $F_5^4(0.05) = 5.19$.

$f < F_5^4(0.05)$, so we accept H_0, that the variances are equal.

Form the pooled variance:

$$s^2 \;=\; (350.8 + 182.0)/(5+6-2) \;=\; 59.2 .$$
$$s \;=\; 7.694 .$$
$$t \;=\; (161.8 - 181.0)/\sqrt{(59.2(1/5 + 1/6))} \;=\; -4.121 .$$

116

From Table 2, the critical t value for 9 degrees of freedom and α = 0.05 is 2.26.

$t < -2.26$, so we accept H_1. Therefore, we conclude that the production lines do not fill the cans equally.

In lots of cases, this is the best test for a difference between the means of two samples. However, in certain special cases, a more powerful test can be used.

7.2.2   MATCHED PAIRS T-TEST

It is often the case that we are interested in determining whether there is some difference in our results due to some treatment which we can apply. However, the effect of the treatment may be obscured by variations in the results caused by other factors.

For example, cloth may be being produced on a long roll, and we wish to test whether or not a certain treatment improves its wear resistance. Unfortunately, wear resistance varies markedly according to position on the roll.

In general, because of outside factors which influence the results, the variance of the data is increased, which reduces the value of the t statistic and makes it more likely that we will accept H_0 rather than H_1. We are less likely to detect a real difference if one exists, and thus the power of the test is reduced. Ideally, we should like to eliminate as far as possible the effects of the outside factors.

A method of reducing these effects is to take pairs of measurements, one with and one without the treatment of interest, matched together as far as possible so as to compensate for the other factors.

Suppose we have n matched pairs of readings:

$x_{11}, x_{12}, \dots x_{1n}$ with the treatment, and
$x_{21}, x_{22}, \dots x_{2n}$ without the treatment,

and x_{1i} is matched with x_{2i} (i.e. measured under the same conditions etc.), then we shall concern ourselves with the

$$\textit{differences:} \qquad d_i \;=\; x_{1i} - x_{2i} \;,\; i = 1, \,..\, n \,.$$

These n values will measure the effect of the treatment of interest, and can be used in a simple t-test to test the hypotheses:

$$H_0: \mu = 0 \quad \text{vs.} \quad H_1: \mu \neq 0 \,,$$

where μ is the mean difference due to the treatment.

Example: Samples are taken at 5 different places on the roll of cloth, and cut in half. One half is treated, and the other not. Measurements of wear resistance (arbitrary units):

Sample:	1	2	3	4	5
Treated:	101	153	86	127	99
Untreated:	95	141	78	119	95
d_i:	6	12	8	8	4

$\bar{d} = 7.6$ and $s^2 = 8.8$. $s = 2.966$, and
$t = (\bar{d}-0)/(s/\sqrt{n}) = 5.73$.

From Table 2, the critical t value with 4 degrees of freedom and $\alpha = 0.05$ is 2.78. Our calculated t value is greater than this. So we accept $H_1: \mu \neq 0$, and assume that the treatment does improve the wear resistance of the cloth.

Note that a matched pairs t-test is only appropriate in these special circumstances, where there is a direct relationship between the values in the two samples. It might also be interesting to carry out a two-sample t-test on the same data.

$$\text{Treated:} \quad \bar{x}_1 = 113.2 \,, \quad s_1^2 = 716.2 \,.$$
$$\text{Untreated:} \quad \bar{x}_2 = 105.6 \,, \quad s_2^2 = 604.8 \,.$$
$$\text{Test variances:} \quad f = s_1^2/s_2^2 = 1.184 \,.$$

The critical value $F_4^4(0.05) = 6.39$, so we accept that the variances are equal.

$$\text{Pooled variance} \quad s^2 = 660.5 \,.$$
$$t = 0.468 \,.$$

This is clearly not significant, and we would accept H_0, that there is no difference between the samples. This illustrates the extra power of the matched pairs t-test in this case. It has reduced the variance from 660.5 to 8.8, and thus increased the value of the t statistic and the power of the test.

	Two Sample	Matched Pairs
Assumptions:	Normal distribution, equal variances, random samples.	Normal distribution, matched samples.
s^2	$\dfrac{(n_1-1)s_1^2 + (n_2-1)s_2^2}{n_1 + n_2 - 2}$	$\dfrac{\sum\limits_{i=1}^{n}(d_i-\overline{d})^2}{n-1}$
t	$\dfrac{\overline{x}_1 - \overline{x}_2}{\sqrt{\left(s^2\left(\frac{1}{n_1} + \frac{1}{n_2}\right)\right)}}$	$\dfrac{\overline{d}}{s/\sqrt{n}}$
Degrees of freedom:	$n_1 + n_2 - 2$	$n - 1$
Preliminary tests:	f-test for equal variances.	None.

7.3 Exercises

1. Tea-breaks in a factory should average 15 minutes in
    length. A foreman suspects that they are actually taking
    longer than this. He times a random sample and finds the
    following durations:
        12, 22, 16, 18, 25, 19 and 21 minutes.
    Are his suspicions correct ?

2. It is suspected that football players have a lower than
    average intelligence. A random sample yields the
    following IQ values:
        97, 83, 74, 91, 63, 85, 88, 83.
    Assuming normal is 100, what conclusions do you draw from
    this ?

3. Two zoos keep hippopotami. In a study over several years, the lifetimes of the animals in the two zoos have been measured, in years:

Zoo A: 5.5, 6.4, 3.8, 4.9
Zoo B: 2.6, 1.9, 3.3, 4.1, 3.8, 2.5 .

Analyse these figures and show if one zoo has a longer life expectancy for its hippos than the other.

4. To test if learning statistics improves IQ, 7 pairs of identical twins were investigated. One twin of each pair was taught statistics, and the other was not.

With statistics: 101   86   112   93   108   125   93
Without statistics: 95   78   103   85   102   115   91

Analyse the results of this experiment.

5. An oil company wishes to test the effectiveness of a new technique for distilling gasoline from crude oil, and compares the percentage gasoline produced with 8 different crudes against the standard method. The results are:

Old method: 32.8 21.0 15.3 18.9 35.7 26.4 20.5 15.3
New method: 33.4 21.8 17.3 20.1 37.2 27.1 21.0 16.0

Test the hypothesis that the new method changes the percentage gasoline produced.

6. 6 children have had "aggressiveness" measured before and after watching a violent film. These are the results:

Child: 1   2   3   4   5   6
Before: 3.1   6.2   9.3   7.4   5.2   4.8
After: 3.9   7.6   9.6   8.6   6.3   6.0

Use two different tests to analyse these results, and show which is the more powerful.

7.4 Computer Projects

1. Write a computer program to perform a two-sample t-test on input data, including the preliminary f-test.

2. Write a program (or amend the one in the text) to perform a matched pairs t-test. If desired, these two programs could be combined into one general purpose t-testing program, with a suitable user interface.

CHAPTER 8

X² Tests

8.1 Goodness of Fit Tests

8.1.1 <u>INTRODUCTION</u>

Moving away from t-tests, we shall next investigate a slightly more general hypothesis-testing problem - checking to see if a theory or model we have developed ties in with some data we have collected. How well does the theory fit the data, allowing for the fact that there is random error in the data ?

In particular, let us consider the following situation: individuals or objects can be classified into one of m mutually exclusive and exhaustive categories. We have developed a theory which implies that the probability of an individual or object being in category i is p_i. We take a random sample of N individuals and discover to which categories they belong, so that n_i individuals are found to be in category i. Is this data in agreement with our theory ?

Define e_i = expected number in category i according to

the theory $= Np_i$.

  We would not really expect n_i to be exactly equal to e_i for all i, because of random errors in the data. We want to be able to test to see if the differences between the n_i and e_i values are big enough to cast doubt on our theory.

Example: The occurrence of girls in families of 4 children.

Theory: If you have a family of 4 children, then X, the number of girls, is Binomially distributed with $p = \frac{1}{2}$.

  In this experiment, "individuals" corresponds to families and there are 5 categories, corresponding to the possible numbers of girls in a family.

i:	0	1	2	3	4
p_i:	$\frac{1}{16}$	$\frac{1}{4}$	$\frac{3}{8}$	$\frac{1}{4}$	$\frac{1}{16}$

We take a sample of 128 families of 4 children, and find the following results:

i:	0	1	2	3	4
n_i:	20	40	40	18	10

From our theory we can calculate the expected numbers $e_i = Np_i$.

i:	0	1	2	3	4
e_i:	8	32	48	32	8

We wish to test the following hypotheses:

  H_0: Theory fits the data

vs. H_1: Theory does not fit the data.

  The statistic used to carry out this test is called χ^2 (chi-squared), and is calculated (approximately) by

$$\chi^2 = \sum_{i=1}^{m} (n_i - e_i)^2/e_i \quad \text{(Formula I)} .$$

It is clear that if H_0 is true, χ^2 will tend to be small, and that if H_1 is true χ^2 will tend to be larger. To get a critical value for the statistic, we make use of the χ^2 distribution which we met in chapter 5. However, the shape of the distribution depends on the number of degrees of freedom, and we need to determine what this should be.

8.1.2 DEGREES OF FREEDOM

The concept of "degrees of freedom" arises in many places in statistics, and it is difficult to give a precise definition, although it is so important. One way of thinking about it is as a measure of the amount of information available to us at any point. That is to say, if we collect N items of data that gives N pieces of information or degrees of freedom to start with. Each time we estimate a model parameter from the data, we "use up" a degree of freedom and reduce the amount of available information for further work.

For example, if we compute the mean $\bar{x}$ of the data, as we usually do in t-tests, that leaves $N-1$ degrees of freedom for further testing. If the model has k parameters to be estimated from the data, we are left with $N-k$ degrees of freedom and we can use these to test the effectiveness of the model. If $k \geq N$, then the model is worthless and there is no way of testing it. This is equivalent to fitting a 5th order polynomial to 3 data points. It can be done, and the model fits the data exactly, but it is meaningless and leads to no insights into the data. Unfortunately, it is often the case that over-complex models are fitted to insufficient data, and used to justify worthless conclusions. So accounting for degrees of freedom in model-fitting is as important as accounting for money spent in balancing a budget.

Returning to the situation where we are given the numbers of individuals in m different categories, we will initially have m degrees of freedom from our data. However, knowing that the total number of individuals must sum to N leaves us with just $m-1$ degrees of freedom. Another way of seeing this is to think that we can freely choose the numbers in $m-1$ of the categories, but this then specifies how many are in the last category to make the total correct.

If the model we are testing requires no parameters to be estimated from the data, then we have $m-1$ degrees of freedom for testing the model and the critical value for χ^2

will be drawn from the tables with $m-1$ degrees of freedom. If however we need to estimate k parameters from the data, then the degrees of freedom remaining are $m-k-1$. Clearly, no model requiring more than $m-1$ parameters can be sensibly fitted.

8.1.3 χ^2 TEST OF MODEL

For our previous example, using the approximate formula I:

$$\chi^2 = \sum_{i=1}^{5} (n_i - e_i)^2 / e_i = \frac{144}{8} + \frac{64}{32} + \frac{64}{48} + \frac{196}{32} + \frac{4}{8}$$

$$= 27.96 .$$

Degrees of freedom $= 5 - 1 = 4$.

From Table 3 with 4 degrees of freedom and $\alpha = 0.05$ we get a critical value of $\chi^2 = 9.488$.

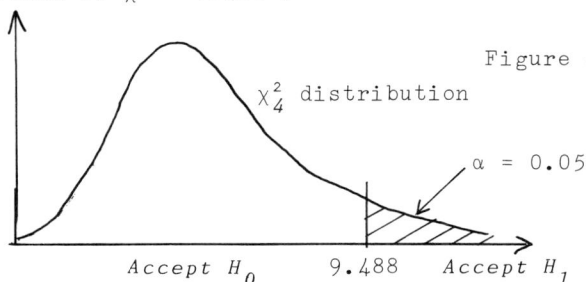

Figure 8.1

χ^2_4 distribution

$\alpha = 0.05$

Accept H_0   9.488   *Accept H_1*

So we accept H_1: the model does not fit the data.

The formula we have used so far for χ^2 (formula I) is only approximate; it is an approximation to the more exact formula:

$$\chi^2 = \sum_{i=1}^{m} 2n_i \, \ell n(n_i / e_i) \quad \text{(Formula II)} .$$

If all the values of n_i and e_i are greater than about 5, then the two formulae will be in general agreement. However, if this is not the case then Formula II should be used. Formula I is often used for hand calculations, because it is simpler, but Formula II should always be preferred for computer programs. Note however that if any of the n_i or e_i

values are zero, then Formula II will clearly fail. In this case it is normal practice to amalgamate categories to ensure that each category has at least one member.

Returning to our example, we have decided that the simple Binomial model with $p = \frac{1}{2}$ does not fit the data. We need to produce a revised theory.

Revised Theory: In a family of 4 children, X, the number of girls, is Binomially distributed with an unknown probability p.

The value of p must be estimated from the data:

$$\hat{p} = \frac{\text{Total number of girls}}{\text{Total number of children}}$$

$$= \frac{1 \times 40 + 2 \times 40 + 3 \times 18 + 4 \times 10}{4 \times 128} = 0.418 .$$

Hence we can find the expected numbers:

i:	0	1	2	3	4
p_i:	0.115	0.330	0.355	0.170	0.030
e_i:	14.7	42.2	45.4	21.8	3.9

Experimental values:

i:	0	1	2	3	4
n_i:	20	40	40	18	10

Note that it is important that the n_i and e_i values sum to the same total - it is usually worth checking for rounding errors when the e_i values have been calculated.

$$\text{Formula I :} \quad \chi^2 = 12.28$$
$$\text{Formula II:} \quad \chi^2 = 9.84 .$$

There is quite a big difference here, and we shall use the more accurate value of 9.84 . One degree of freedom was used up in estimating p, leaving 3 degrees of freedom for the χ^2 test.

$$\text{Critical values:} \quad \alpha = 0.05: 7.815$$
$$\alpha = 0.01: 11.34 .$$

Since $\chi^2 = 9.84$, we may either accept or reject H_0, depending on the level of significance chosen.

8.2 Contingency Tables

The general idea of the goodness of fit test can be extended to an important special case. Suppose we have two sets of categories, based on two factors A and B, where A has n_a categories and B has n_b categories. N individuals are classified according to both factors, and n_{ij} is the number of individuals in category i for A and category j for B.

Assume that the theory we wish to test is that classification according to factor A has no influence on classification according to factor B. In other words, A and B are independent.

H_0: A and B are independent

vs. H_1: A and B are not independent.

We need to estimate the following parameters of our model:

$$p_i^a = \text{P(Classified in category } i \text{ for A)} \quad i = 1, \ldots n_a$$
$$p_j^b = \text{P(Classified in category } j \text{ for B)} \quad j = 1, \ldots n_b .$$

If H_0 is true,

$$\text{P(Classified in } i \text{ for A and } j \text{ for B)} = p_i^a p_j^b .$$

So the expected number $e_{ij} = N p_i^a p_j^b .$

We shall estimate these parameters as follows:

$$\hat{p}_i^a = \frac{1}{N} \sum_{j=1}^{n_b} n_{ij} ; \quad \hat{p}_j^b = \frac{1}{N} \sum_{i=1}^{n_a} n_{ij} .$$

The number of degrees of freedom used up in estimating the probabilities for factor A in $n_a - 1$. This is because we know that they must sum to 1, so all the probabilities are known once $n_a - 1$ of them have been found. Similarly, the number of degrees of freedom for factor B is $n_b - 1$. At the start, the total degrees of freedom are $n_a n_b - 1$. So the number of degrees of freedom left for testing

$$= n_a n_b - 1 - (n_a - 1) - (n_b - 1)$$
$$= (n_a - 1)(n_b - 1) .$$

Example: A survey has been carried out on 100 children, investigating the relationship between TV watching and school performance. Each child's TV watching has been classified as little, moderate, or a lot, and school performance has been classified as poor, moderate, or good. The numbers of children in the various categories are:

		TV Watching		
		Little	Moderate	A lot
	Poor	5	10	10
School	Moderate	10	20	15
Performance	Good	15	10	5

We wish to test:

H_0: TV watching and school performance are independent,

vs. H_1: They are related.

If H_0 is true, we can estimate the following probabilities:

P(School performance poor) $\quad = \hat{p}_1^a \quad = \dfrac{5+10+10}{100} = 0.25$

P(School performance moderate) $\quad = \hat{p}_2^a \quad = \dfrac{10+20+15}{100} = 0.45$

P(School performance good) $\quad = \hat{p}_3^a \quad = \dfrac{15+10+5}{100} = 0.30$.

Similarly, for TV watching:

$$\hat{p}_1^b = 0.30; \quad \hat{p}_2^b = 0.40; \quad \hat{p}_3^b = 0.30 .$$

Assuming independence,

P(School performance poor and TV watching little)

$= \hat{p}_1^a \hat{p}_1^b = 0.25 \times 0.30 = 0.075$.

Expected number in this group $= 100 \times 0.075 = 7.5$.

We can carry out this calculation for each cell in the table, and lay out the results in a *contingency table*. For each cell, the unbracketed value denotes the observed number in the group (n_{ij}), and the bracketed value the expected number (e_{ij}), assuming independence between the two factors.

		TV Watching		
		Little	Moderate	A Lot
School Performance	Poor	5 (7.5)	10 (10)	10 (7.5)
	Moderate	10 (13.5)	20 (18)	15 (13.5)
	Good	15 (9)	10 (12)	5 (9)

Note that it is necessary for the n_{ij} and e_{ij} values to sum to the same total in each row and column.

The value of χ^2 is given by

$$\chi^2 = \sum_i \sum_j 2n_{ij} \ln(n_{ij}/e_{ij}) \qquad \text{(Formula II)}$$

$$= 8.87 .$$

The number of degrees of freedom = (3-1)(3-1) = 4. Another way of seeing this is to say that once 4 of the e_{ij} values have been filled in, the rest follow automatically because of the requirements on the row and column sums.

The critical χ^2 with 4 degrees of freedom and $\alpha = 0.05$ is 9.488. Therefore, we accept H_0, that the two factors are independent. In practice, since the test is nearly significant, we might suspect it is not pwerful enough, and repeat the experiment with more children.

8.3 Exercises

1. 300 students were classified according to the size of school they came from and according to their performance in the first year exams at university.

	School size		
	Small	Medium	Large
Class 1 or 2.1	18	51	46
Class 2.2 and below	42	79	64

    Test for a relationship between size of school and first year performance.

2. 120 litters, each of 3 guinea pigs, were investigated, and the following numbers of females were found:

No. of females	No. of litters
0	12
1	50
2	41
3	17

Set up and test a plausible statistical model for the occurrence of female guinea pigs.

3. The following table gives the number of phone calls to a switchboard during 100 5-minute intervals:

Number of calls:	0	1	2	3	4	5	6	7
Frequency:	5	22	26	23	11	7	4	2

a) Does this data fit a Poisson distribution with mean 2 ?
b) Does it fit a Poisson distribution at all ?

4. The exam grades for 50 students are given below. Are these grades Normally distributed ?

Exam Grades	No. of students
11-25	3
26-40	8
41-55	22
56-70	11
71-85	5
86-100	1

5. Probation officers have been asked to rank 500 criminal offenders on a scale 1 to 5, according to their likelihood of being reconvicted after completing their sentence (1 = low risk, 5 = high risk). The numbers reconvicted within one year of completing their sentence have been noted, with the following results:

Ranking	No. Reconvicted	No. not reconvicted
1	34	46
2	100	140
3	39	81
4	19	21
5	8	12

Is there any evidence that the probation officers'
ranking can be used to predict an offender's chance of
reconviction ?

8.4 Computer Projects

1. Write a program to test a Binomial model for the
    occurrence of individuals in *n* categories. It should
    either allow the user to input a predetermined value of *p*,
    or estimate *p* from the data.

2. Write a program to carry out a contingency table test for
    the independence of two factors.

    (For both these you will want to look up tables of critical
    χ^2 values - a slight variant of the procedure in chapter 7
    for t tables will enable you to do this).

CHAPTER 9

Linear Regression

9.1 Introduction

Linear regression is one way in which a possible relationship between two variables can be explored. For example, suppose we were interested in the relationship between height and weight for a particular group of individuals. We might measure both the heights and weights of a number of individuals, and wish to use this data to explore how weight depends on height.

In general, suppose we have two variables, X and Y, which can be measured simultaneously on a set of N different individuals or objects. We can plot this data as a set of N points on a two-dimensional graph.

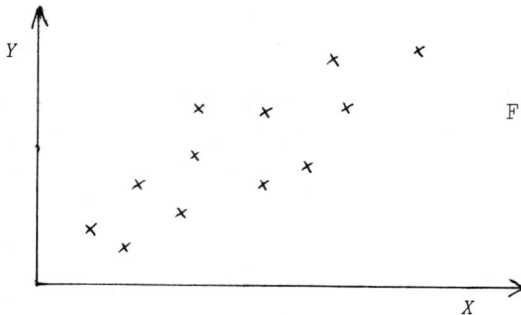

Figure 9.1

One possibility is to model the relationship between X and Y by a straight line on the graph, passing through the cloud of data points. But where should we draw this line so as to pass as close as possible to all the data points? Putting this another way, we wish to fit a *linear model* of the form:

$$Y = a + bX$$

to the data, where a is the intercept of the line on the Y axis and b is the slope of the line. If $y_1, y_2, \ldots y_N$ is the set of Y values and $x_1, x_2, \ldots x_N$ the corresponding set of X values, then we assume:

$$y_i = a + bx_i + \varepsilon_i, \quad i = 1, \ldots N,$$

where ε_i is the difference between the ith data value and the model, and the ε_i values are assumed to be Normally distributed with mean 0.

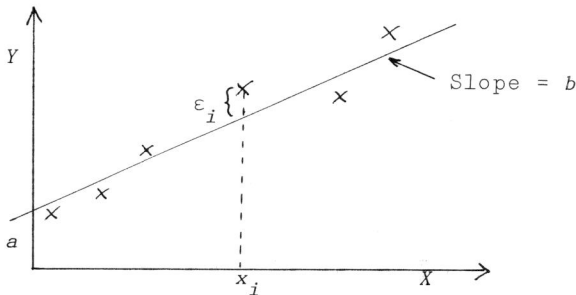

Figure 9.2

Notice the important difference between the variables X and Y. All the errors in fitting the line to the data are assumed to be errors in Y - X is assumed to be fixed. We call Y the *dependent* variable and X the *independent* variable.

The dependent variable (usually called Y) is the variable whose values we wish to predict from known X values. We assume the errors are errors in predicting Y. The independent variable (usually called X) is assumed to be the one from which values of Y are to be predicted. We say that we "regress Y on X".

It is possible to regress X on Y, and this will give a different line, predicting X values from known Y values.

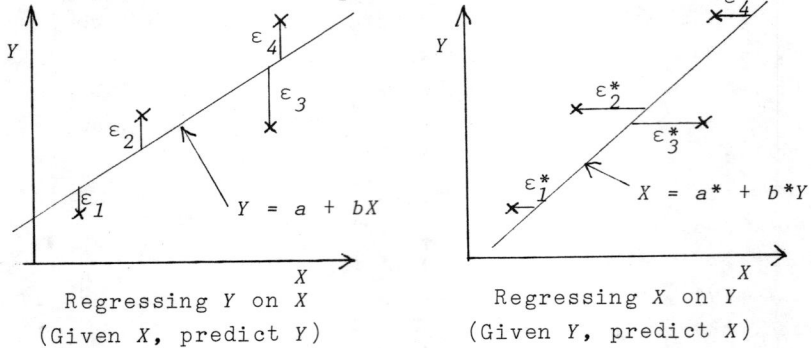

Figure 9.3

Regressing Y on X
(Given X, predict Y)

Regressing X on Y
(Given Y, predict X)

Example: Measurements have been made of the heights and weights of 6 people, as follows:

Height (inches)	Weight (pounds)
61	106
67	138
73	173
72	218
65	165
70	160

We wish to find an equation which will predict an individual's weight given their height. Plot the data:

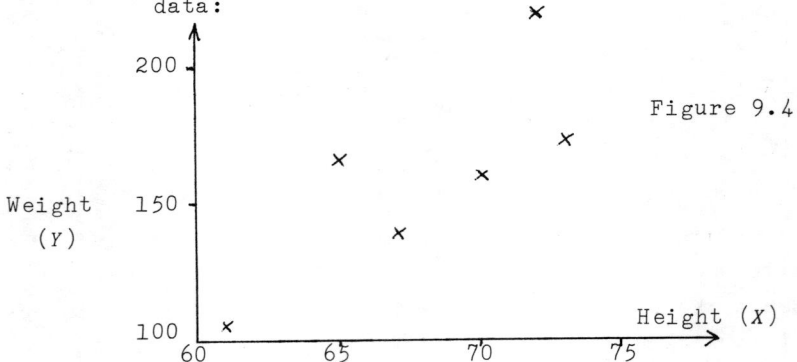

Figure 9.4

Regress Y (Weight) against X (Height) - i.e. find a line $Y = a + bX$ which "best fits" the plotted data points. The problem thus becomes that of finding a and b which best fit the data. First we need to define the criterion for what we mean by "best fit".

134

9.2 "Best Fit" and Regression Formulae

Suppose a and b have arbitrary values, and from the line they define we predict the Y value for each X value x_i:

$$\hat{y}_i = a + bx_i .$$

The criterion we shall use to find the best values of a and b is that of *least squared error*. Note that this is not the only possible criterion - many others can be used, but this one leads to nice mathematical formulae.

Define $\quad E = \sum_{i=1}^{N} (y_i - \hat{y}_i)^2 .$

This is the total squared difference between the true and predicted Y values. Choose a and b so as to minimise E.

Rewrite: $\quad E = \sum_{i=1}^{N} (y_i - (a + bx_i))^2$

$$= \sum_{i=1}^{N} (y_i^2 - 2y_i(a + bx_i) + (a + bx_i)^2)$$

$$= \sum_{i=1}^{N} y_i^2 - 2a \sum_{i=1}^{N} y_i - 2b \sum_{i=1}^{N} x_i y_i + \sum_{i=1}^{N} (a^2 + 2abx_i + b^2 x_i^2)$$

$$= \sum_{i=1}^{N} y_i^2 - 2Na\overline{y} - 2b \sum_{i=1}^{N} x_i y_i + Na^2 + 2Nab\overline{x} + b^2 \sum_{i=1}^{N} x_i^2 ,$$

where $\quad \overline{x} = \dfrac{1}{N} \sum_{i=1}^{N} x_i \quad$ and $\quad \overline{y} = \dfrac{1}{N} \sum_{i=1}^{N} y_i .$

To find values of a and b which minimise E, we shall differentiate it with respect to both a and b in turn, holding the other constant (this is *partial differentiation*).

$$\frac{\partial E}{\partial a} = -2N\overline{y} + 2Na + 2Nb\overline{x} = 0$$

$$\Rightarrow \quad \underline{a = \overline{y} - b\overline{x}} .$$

$$\frac{\partial E}{\partial b} = -2 \sum_{i=1}^{N} x_i y_i + 2Na\overline{x} + 2b \sum_{i=1}^{N} x_i^2$$

135

$$= 2 \sum_{i=1}^{N} x_i y_i + 2N\overline{x}(\overline{y} - b\overline{x}) + 2b \sum_{i=1}^{N} x_i^2 = 0 \ .$$

$$\Rightarrow \quad b = \frac{\sum_{i=1}^{N} x_i y_i - N\overline{xy}}{\sum_{i=1}^{N} x_i^2 - N\overline{x}^2} \qquad (I) \ .$$

We can call this Formula I for b. There is an alternative formula, as we can see by studying the following expressions:

$$\sum_{i=1}^{N} (x_i - \overline{x})^2 = \sum_{i=1}^{N} x_i^2 - 2\overline{x} \sum_{i=1}^{N} x_i + N\overline{x}^2$$

$$= \sum_{i=1}^{N} x_i^2 - N\overline{x}^2$$

$$= \text{Denominator of Formula I} \ .$$

Also $\displaystyle\sum_{i=1}^{N} (x_i - \overline{x})(y_i - \overline{y}) = \sum_{i=1}^{N} x_i y_i - \overline{x} \sum_{i=1}^{N} y_i - \overline{y} \sum_{i=1}^{N} x_i + N\overline{xy}$

$$= \sum_{i=1}^{N} x_i y_i - N\overline{xy}$$

$$= \text{Numerator of Formula I} \ .$$

So Formula II is:

$$b = \frac{\sum_{i=1}^{N} (x_i - \overline{x})(y_i - \overline{y})}{\sum_{i=1}^{N} (x_i - \overline{x})^2} \qquad (II) \ .$$

From now on, we shall adopt a form of shorthand notation:

Let $S_{XX} = \displaystyle\sum_{i=1}^{N} (x_i - \overline{x})^2 = \sum_{i=1}^{N} x_i^2 - N\overline{x}^2$,

$S_{XY} = \displaystyle\sum_{i=1}^{N} (x_i - \overline{x})(y_i - \overline{y}) = \sum_{i=1}^{N} x_i y_i - N\overline{xy}$,

$S_{YY} = \displaystyle\sum_{i=1}^{N} (y_i - \overline{y})^2 = \sum_{i=1}^{N} y_i^2 - N\overline{y}^2$.

Then

$$b = S_{XY}/S_{XX} \quad \text{and} \quad a = \overline{y} - b\overline{x} \ .$$

The two alternative formulae give the same results mathematically, but computationally they are to be preferred under the following circumstances:

Formula I - for hand calculations with a small set of data, especially if $\bar{x}$ or $\bar{y}$ are not simple integers. Then subtracting them from the data will lead to more complex calculations with more decimals.

Formula II - for computer calculations. Subtracting $\bar{x}$ and $\bar{y}$ beforehand reduces the size of the numbers to be squared, and leads to smaller rounding errors.

9.2.1 NUMERICAL CALCULATIONS ON EXAMPLE DATA

By Formula I -

x_i	y_i	x_i^2	$x_i y_i$
61	106	3721	6466
67	138	4489	9246
73	173	5329	12629
72	218	5184	15696
65	165	4225	10725
70	160	4900	11200
Totals 408	960	27848	65962

$N = 6$; $\bar{x} = 408/6 = 68$; $\bar{y} = 960/6 = 160$.

$S_{XX} = 27848 - 6 \times 68^2 = 104$.

$S_{XY} = 65962 - 6 \times 68 \times 160 = 682$.

By Formula II -

x_i	y_i	$x_i - \bar{x}$	$(x_i - \bar{x})^2$	$y_i - \bar{y}$	$(x_i - \bar{x})(y_i - \bar{y})$
61	106	-7	49	-54	378
67	138	-1	1	-22	22
73	173	5	25	13	65
72	218	4	16	58	232
65	165	-3	9	5	-15
70	160	2	4	0	0
Total 408	960	0	104	0	682

$S_{XX} = 104$; $S_{XY} = 682$.

(In practice, we would only use one or other method of calculating S_{XX} and S_{XY}, not both. We have only done so here to show that the same result is obtained by either route).

Best slope: $b = S_{XY}/S_{XX} = 682/104 = 6.5577$.

$$a = \overline{y} - b\overline{x} = 160 - 6.5577 \times 68 = -285.923 .$$

So the straight line that best fits the data is

$$Y = -285.923 + 6.5577\ X .$$

Warning

This straight line model is fine as far as it goes. However, it has only been fitted to a small set of data over a limited range, and it would clearly be gross folly to use it to extrapolate much outside that range. For example, for a man 80 inches tall, the predicted weight is

$$Y = -285.923 + 6.5577 \times 80 = 238.69,$$

which is not totally unreasonable. However, our formula predicts that someone 40 inches tall weighs

$$Y = -285.923 + 6.5577 \times 40 = -23.615,$$

which is clearly ridiculous. This shows that any mathematical model, however carefully fitted to data, needs to be treated with caution and respect and not expected to give more information than is really available.

9.3 Variances of Regression Estimates

So far we have concentrated on estimating a and b, and thus on fitting the regression line to the data. As important, or perhaps more so, are two further questions which should be resolved.

1. What is the error in fitting any point ? In other words, if $\varepsilon_i = y_i - \hat{y}_i$,

   what is the variance of the ε_i values ? (We can use this to test whether a new data point belongs to the same data set and can be fitted by the same line, or should be treated as an "outlier" or foreign data point).

2. What is the uncertainty in fitting the line itself ? In other words, what is the uncertainty in estimating the slope b ? This can be used to test whether the slope really could just as well be taken to be equal to zero (i.e. there is no relationship between X and Y). Hence we may critically investigate how meaningful our fitted relationship really is.

9.3.1   UNCERTAINTY IN FITTING DATA POINTS

We can estimate the variance of the residual errors ε_i in the Y values by calculating

$$s_e^2 = (S_{YY} - b^2 S_{XX})/(N-2) \ .$$

To see where this comes from, note that

$$y_i = \varepsilon_i + (a + bx_i) \ .$$

So $\quad \text{Var}(y_i) = \text{Var}(\varepsilon_i) + \text{Var}(a + bx_i)$

$$= \text{Var}(\varepsilon_i) + b^2 \text{Var}(x_i) \ ,$$

$$\Rightarrow \quad \text{Var}(\varepsilon_i) = \text{Var}(y_i) - b^2 \text{Var}(x_i) \ .$$

Now, once we have estimated $\overline{y}$ and b, there are $N-2$ degrees of freedom left for estimating variances. So:

$$\text{Var}(y_i) \text{ is estimated by } S_{YY}/(N-2) \ ,$$
$$\text{Var}(x_i) \text{ is estimated by } S_{XX}/(N-2) \ ,$$

and $\quad \text{Var}(\varepsilon_i)$ is estimated by s_e^2 , given by the formula above.

For the example data set, $S_{YY} = 6958$. So $s_e^2 = (6958 - 6.5577^2 \times 104)/4 = 621.41$, and $s_e = 24.928$.

s_e is the standard deviation of the residual errors, and gives a measure of the spread of the data about the regression line. Suppose we are given a newpoint (x^*, y^*) and want to test to see if it could belong to the same data set that was used to fit the line. We can use our value of s_e to carry out a t-test.

H_0: New point is fitted by the line -
$$y^* = a + bx^* + \varepsilon^*, \text{ where } \varepsilon^* \text{ has mean } 0.$$

H_1: The mean of ε^* is not 0.

        Form $t = (y^* - (a + bx^*))/s_e$.

If H_0 is true, this has a t-distribution with $N-2$ degrees of freedom.

Example: A new individual has height 71 inches and weight 110 lbs. Could this individual be fitted by the same regression line as the others ?

    $x^* = 71$, $y^* = 110$.

    $t = (110 - (-285.923 + 6.5577 \times 71))/24.928$

      $= -2.795$.

    Critical $t_4(0.05) = 2.78$, from Table 2.

We therefore accept H_1 (only just) and decide that this person is not a member of the same group as the others, and is not fitted by the same line.

9.3.2 <u>UNCERTAINTY IN THE SLOPE</u>

      The value of the slope b that we have computed is a random variable, since it depends on the data which is subject to random errors. We may estimate the variance of b by

$$s_b^2 = s_e^2/S_{XX} .$$

Knowing this, we can test the hypothesis that there is really no relationship between X and Y, i.e. the slope is zero.

    H_0: $b = 0$ vs. H_1: $b \neq 0$.

Calculate $t = (b - 0)/s_b = b\sqrt{S_{XX}}/s_e$.

Example: Using our current example data,

    $s_b^2 = s_e^2/S_{XX} = 621.41/104 = 5.975$.

    $s_b = 2.444$,

and $t = 6.5577/2.444 = 2.683$.

The critical $t_4(0.05) = 2.78$, and so we accept H_0 (just about), that the slope is really zero and there is no relationship between X and Y. In practice, in order to show a real relationship between X and Y, we would need a more powerful test. This could be obtained most simply by collecting more data.

9.4 Non-Linear Transformations

One of the interesting features of linear regression is that we can sometimes use it to fit models to data which are not linear at all. For example, suppose we wanted to fit the model:

$$Y = ab^X .$$

Y might be some measure of growth, such as a cost of living index, while X is a time variable. This model says that Y is increasing by a common multiplicative factor each time period. But this model is not linear, and we cannot use linear regression on it as it stands. So we apply one of the fundamental principles of mathematical modelling: "If you have a problem you cannot solve, change it into a problem you can solve". In this case, we make the change by means of a *non-linear transformation*, i.e. taking logs:

$$log Y = log a + X log b .$$

Now we must make the assumption that the errors in Y are multiplicative, so that if (x_i, y_i) is a data point we can write:

$$log y_i = log a + x_i log b + \varepsilon_i ,$$

where the ε_i values are independently distributed with mean zero. If this is reasonable, then we can use linear regression to fit the model, by defining the following transformed variables:

$$Y' = log Y \qquad\qquad a' = log a$$
$$X' = X \qquad\qquad b' = log b .$$

The model now is:

$$Y' = a' + b'X',$$

and this can be fitted by ordinary linear regression.

Example:

Year (X)	Cost of living Index (Y)
1	102
2	108
3	117
4	125
5	136
6	150

Taking logs to base 10, we get:

x_i	y_i'	x_i^2	$x_i y_i'$
1	2.009	1	2.009
2	2.033	4	4.066
3	2.068	9	6.204
4	2.097	16	8.388
5	2.134	25	10.670
6	2.176	36	13.056
Totals 21	12.517	91	44.393

$\bar{x} = 3.5$; $\bar{y} = 2.086$.

$S_{XX} = 91 - 6 \times 3.5^2 = 17.5$.

$S_{XY} = 44.393 - 6 \times 3.5 \times 2.086 = 0.5835$.

Therefore, $b' = S_{XY}/S_{XX} = 0.03334$.

$a' = 2.086 - 0.03334 \times 3.5 = 1.9693$.

So the line fitted to the transformed data is:

$$Y' = 1.9693 + 0.03334X' .$$

Transforming back to the original model, we have:

$$a = 10^{a'} = 93.175; \quad b = 10^{b'} = 1.0798 .$$

$$\underline{Y = 93.175 \times (1.0798)^X .}$$

This can be interpreted as meaning that the average yearly inflation factor is 1.0798, or about 8%.

This is not the only non-linear model that can be fitted in this way. Some other examples follow, and many more can be conceived.

a) $$Y = aX^b .$$

Transformation - take logs:

$X' = logX$ $a' = loga$

$Y' = logY$ $b' = b$.

b)
$$Y = 1/(a + bX) \ .$$

Transformation - take reciprocals:

$X' = X$ $a' = a$

$Y' = 1/Y$ $b' = b \ .$

c)
$$Y = a + b sinX \ .$$

Transformation - just change X:

$X' = sinX$ $a' = a$

$Y' = Y$ $b' = b \ .$

9.5 Exercises

1. An oil company wishes to estimate the depth of the oil/ water contact in a newly-discovered oilfield. To do this, pressures have been measured at various depths in both the oil-bearing and water-filled portions of the reservoir. The data is as follows:

OIL		WATER	
Depth (ft)	Pressure (psi)	Depth (ft)	Pressure (psi)
8125	4260	9650	4800
8200	4200	9800	4860
8300	4330	9900	5000
8500	4370	9980	5085
8660	4375	10170	5100
8770	4500	10200	5235
8900	4450		
9040	4490		
9160	4590		
9325	4550		

The depths are accurately known, but there are errors in the pressures due to gauge problems. Fit a straight line to each set of data, and find the intersection of the lines, which will give the desired oil/water contact.

2. Measurements of X and Y have been made:

X: 2   6   7   9   10   11

Y: 3   7   13   11   20   26

Fit models of the following forms to this data:

a) $Y = a + bX^2$

b) $Y = aX^b$

143

c) $Y = ab^X$.

3. A zoologist is studying pink-eared wombats from Patagonia. He has a theory that it is possible to estimate the age of an animal by measuring the length of its tail. He makes measurements on his herd of captive wombats, with the following results:

Length of tail (inches)	Age (months)
5	3
7.5	8
2	7
11	9
10	12
1	2
13.5	15.5
15	13
18	13

Regress age on length of tail to find a relationship between them. Test to see if the relationship is real. Someone else sends data about their wombat, which is 4 months old and has a tail 10 inches long. Is this likely to be a pink-eared Patagonian wombat ? Repeat all these calculations, regressing length of tail on age.

4. Fit a model of the form $Y = \sqrt[a]{bX}$ $(= (bX)^{1/a})$ to the following data:

X:	1	10	1000	10,000
Y:	12.5892	31.6228	125.8925	316.2278

A new point is measured with values (3162.23, 398.1072). Test to see if your model also fits this new point.

9.6 Computer Projects

1. Write a program to carry out simple linear regression, including testing for the reality of the relationship.

2. Extend your program so that the user has available a number of possible transformations for fitting non-linear models to the data.

CHAPTER 10

Analysis of Variance

10.1 Basic Concepts

Analysis of Variance (or ANOVA as it is commonly abbreviated) is a very important technique for testing whether models which have been fitted to data really help to understand the data, or are totally spurious. We shall consider several applications of the method, but its applicability is much wider than we shall be able to see here.

Let us assume that we have obtained N measurements of some variable Y, which have been obtained under various different experimental conditions. Label the measurements $y_{ijk..}$, with the subscripts relating to the circumstances of the particular experiment. We fit some kind of model to this data, with a number of parameters, and then we want to see if the model really explains anything.

Test H_0: Model spurious (i.e. all parameters = 0)

vs. H_1: Model does explain the data (i.e. not all parameters = 0).

To carry out this test, we shall assume that each data value is made up of 3 parts:

Data value = Overall mean + Model value + Residual error.

The first thing to do is always to get rid of the overall mean, which is removed from the problem as quickly as possible.

145

It is estimated by:
$$\bar{y} = \frac{1}{N} \sum^{N} (\text{Data values}).$$

Then we get:

      Data value $- \bar{y}$ = Model value + Residual error .

The model value describes how much of the data is dependent on the fitted model, and the residual error is what is left over, totally unrelated to the model. If H_0 is true, then the model values can all be taken to be zero. To test H_0, we need a test statistic, and this is generated using *sums of squares*.

Define: Total Sum of Squares $= \sum^{N} (\text{Data value})^2$.

      Corrected Total Sum of Squares $= CTSS$
$$= \sum^{N} (\text{Data value} - \bar{y})^2 .$$
      Model Sum of Squares $= MSS = \sum^{N} (\text{Model value})^2$.
      Residual Sum of Squares $= RSS$
$$= \sum^{N} (\text{Residual error})^2 .$$

Now, the principal result of ANOVA is that, provided the model parameters have been estimated in a sensible way, we find that

$$CTSS = MSS + RSS .$$

Thus the (corrected) total sum of squares, which measures the total variation of the data about the mean, can be divided into two parts: one (MSS) measures the variation in the data "explained" by the model, and the other (RSS) measures the variation left unexplained by the model. The relative sizes of these two quantities are going to be of interest.

Suppose now that the model has m parameters which have been fitted, i.e. m degrees of freedom. The corrected total sum of squares has $N-1$ degrees of freedom (one is lost in estimating the overall mean). The degrees of freedom for the model and the residual error in fact sum to the total degrees of freedom, in exactly the same way as the sums of squares.

$$N-1 = m + N-m-1 .$$
      Total   Model   Residual

If H_0 is true, then the model sum of squares is just a collection of random errors, and MSS will be a χ^2 random variable with m degrees of freedom. We define the "mean model sum of squares" to be equal to MSS/m. Similarly, RSS is a χ^2 random variable with $N-m-1$ degrees of freedom. The "mean residual sum of squares" is $RSS/(N-m-1)$. This is equal to s^2, the estimate of the residual error variance.

Our test statistic is the ratio of these two mean sums of squares: $\qquad f \quad = \quad \dfrac{MSS/m}{RSS/(N-m-1)} \qquad$,

which if H_0 is true has an f-distribution with m degrees of freedom on the top and $N-m-1$ on the bottom.

We look up the critical value $F_{N-m-1}^{m}(\alpha)$ in Table 4, and if the calculated value of f is less than this, we accept H_0. This means that we do not have sufficiently convincing evidence that the model really does explain the data. On the other hand, if f is greater than the critical value, we accept H_1, that the model parameters are not all zero, and that it does explain at least part of the behaviour of the data.

Normally, these calculations are carried out in a standard format, known as an ANOVA table. This has five columns, as follows:

Source of Variation	Degrees of Freedom	Sum of Squares	Mean Sum of Squares	f
Model	m	MSS	MSS/m	
Residual	$N-m-1$	RSS	$RSS/(N-m-1)$	
Total	$N-1$	$CTSS$		

Notes: 1. The second and third columns both must sum to the total degrees of freedom and sum of squares, respectively.
2. The value of RSS is normally found by differencing:
$$RSS \quad = \quad CTSS \quad - \quad MSS \quad .$$

We have given an outline of the basic ideas behind ANOVA; next we shall explore some specific examples.

10.2 ANOVA in Linear Regression

In this case we have N data values $y_1, y_2, \ldots y_N$, each corresponding to a value of the independent variable X. The model is:
$$y_i = a + bx_i + \varepsilon_i .$$
We wish to test:

H_0: Model is spurious ($b = 0$)

vs. H_1: Model explains the data ($b \neq 0$) .

In fact, we have already done this test in chapter 9, using the variance of b and a t-statistic. We shall now see how to do the same test using ANOVA and an f-statistic, and show that the results are the same.

In linear regression, the residual error variance is
$$s_e^2 = (S_{YY} - b^2 S_{XX})/(N-2) .$$
Rewrite this as: $\quad S_{YY} = b^2 S_{XX} + (N-2)s_e^2 .$

Now $\quad S_{YY} = \sum_{i=1}^{N} (y_i - \overline{y})^2 = CTSS$, the corrected total sum of squares,

and $\quad b^2 S_{XX} = b^2 \sum_{i=1}^{N} (x_i - \overline{x})^2 .$

To show that this is equal to the model sum of squares, rewrite the model using $a = \overline{y} - b\overline{x}$:
$$y_i - \overline{y} = b(x_i - \overline{x}) + \varepsilon_i .$$
So ε_i is the residual error and $b(x_i - \overline{x})$ is the model value.
$$MSS = \sum_{i=1}^{N} b^2 (x_i - \overline{x})^2 = b^2 S_{XX} .$$
Thus it is clear that $(N-2)s_e^2 = RSS$.

The model has one degree of freedom, that needed to estimate b (a is given once we know $\overline{x}$, $\overline{y}$ and b). We thus have all the results we need to draw up an ANOVA table to test the significance of the linear regression model.

Example: Use the regression example of chapter 9, of weight against height.

$N = 6$; $S_{XX} = 104$; $S_{YY} = CTSS = 6958$; $b = 6.5577$.

Then $MSS = b^2 S_{XX} = 6.5577^2 \times 104 = 4472.36$.

Source of Variation	Degrees of Freedom	Sum of Squares	Mean Sum of Squares	f
Model	1	4472.36	4472.36	
Residual	4*	2485.64*	621.41	7.20
Total	5	6958.0		

(* - values obtained by differencing).

From Table 4, $F_4^1(0.05) = 7.71$.
This implies that we should accept H_0, or that there is not (quite) sufficient evidence to show that the regression is significant with an error probability of 5%.

Notice that the test statistic

$$f = \frac{MSS/1}{RSS/(N-2)} = b^2 S_{XX}/s_e^2 \quad .$$

In chapter 9, when we tested the hypothesis $b = 0$, we used the statistic $\quad t = b/s_b = b\sqrt{S_{XX}}/s_e$.

It is clear that $f = t^2$, so the two tests are in fact equivalent. Just check that the critical f-value (7.71) is in fact also the square of the critical t-value (2.78).

10.3 Differences between Sample Means

In chapter 7 we saw how to test for a difference between two sample means, by means of two-sample t-tests. But how do we proceed when there are more than two samples ?

Let there be k different samples, each of size n, and let μ_i be the true mean of the random variable X_i from which the ith sample is drawn. We wish to test

H_0: $\mu_1 = \mu_2 = \ldots = \mu_k$
vs. H_1: At least one μ_i is different from the others.

It might be suggested that we should use a series of two-sample t-tests, comparing each pair of samples in turn.

We would have to carry out $^kC_2$ tests, which equals $k(k-1)/2$. For example, if k were equal to 10 we should have to carry out 45 tests. Apart from the effort needed, there is a more fundamental problem. Assume H_0 is true - then the probability that any one test gives the correct result is $1-\alpha = 0.95$, say. But the probability that we will get the correct result on *all* our tests is $(1-\alpha)^{k(k-1)/2}$, which is $(0.95)^{45}$ in our example, or 0.099. This success probability is unacceptably low. We must reject this approach and turn instead to ANOVA.

The model is: jth value from sample i $= \mu_i + \varepsilon_j$.

Or, if μ is the overall mean,

jth value from sample i - overall mean

$$= (\mu_i - \mu) + \varepsilon_j$$
$$= \text{Model value} + \text{Residual error.}$$

Before proceeding with the details of the analysis, it is worth saying a word about notation.

y_{ij} is the jth data value in the ith sample. Each sample contains n data values, so the total amount of data $N = nk$. Whenever we average over a subscript, we shall replace that subscript by a dot:

$$\text{Average value in } i\text{th sample} = \overline{y}_{i.} = \frac{1}{n} \sum_{j=1}^{n} y_{ij} .$$

$$\text{Overall average } \overline{y}_{..} = \frac{1}{N} \sum_{i=1}^{k} \sum_{j=1}^{n} y_{ij} .$$

The Corrected Total Sum of Squares

$$CTSS = \sum_{i=1}^{k} \sum_{j=1}^{n} (y_{ij} - \overline{y})^2 .$$

To show that this breaks into 2 separate parts, write it as:

$$CTSS = \sum_{i=1}^{k} \sum_{j=1}^{n} ((y_{ij} - \overline{y}_{i.}) + (\overline{y}_{i.} - \overline{y}_{..}))^2$$

$$= \sum_{i=1}^{k} \sum_{j=1}^{n} \big[(y_{ij} - \overline{y}_{i.})^2 + 2(y_{ij} - \overline{y}_{i.})(\overline{y}_{i.} - \overline{y}_{..}) + (\overline{y}_{i.} - \overline{y}_{..})^2 \big] .$$

The middle term is $\sum_{i=1}^{k} \sum_{j=1}^{n} (y_{ij} - \overline{y}_{i.})(\overline{y}_{i.} - \overline{y}_{..})$

$$= 2 \sum_{i=1}^{k} (\overline{y}_{i.} - \overline{y}_{..}) \sum_{j=1}^{n} (y_{ij} - \overline{y}_{i.}) ,$$

which is equal to zero, from the fact that $\overline{y}_{i.}$ is the average of the ith sample, and thus $\sum\limits_{j=1}^{n} (y_{ij} - \bar{y}_{i.}) = 0$.

The first term in the expansion of $CTSS$ is

$$\sum_{i=1}^{k} \sum_{j=1}^{n} (\bar{y}_{i.} - \bar{y}_{..})^2 = n \sum_{i=1}^{k} (\bar{y}_{i.} - \bar{y}_{..})^2 .$$

So $CTSS = n \sum\limits_{i=1}^{k} (\bar{y}_{i.} - \bar{y}_{..})^2 + \sum\limits_{i=1}^{k} \sum\limits_{j=1}^{n} (y_{ij} - \bar{y}_{i.})^2 .$

The first term measures the differences due to the different sample means, and can thus be interpreted as the Model Sum of Squares (MSS). It is sometimes known as the "between samples" sum of squares. The second term reflects the differences between the individual data values and the sample means, and is therefore the Residual Sum of Squares (RSS); otherwise known as the "within samples" sum of squares.

The total degrees of freedom are $N-1$ ($= nk-1$). The model has k parameters (the sample means), but once we have estimated $k-1$ of these, the kth one follows automatically from knowing the overall average $\bar{y}_{..}$. So the model has $k-1$ degrees of freedom. Therefore, RSS has $nk-1 - (k-1) = (n-1)k$ degrees of freedom.

Example: 4 production lines are filling cans of peas. To test their consistency, 5 cans are randomly chosen from each line, opened, and the peas are counted. The results are:

Line 1: 57, 63, 72, 59, 52
Line 2: 47, 62, 51, 53, 44
Line 3: 62, 81, 71, 59, 65
Line 4: 50, 64, 44, 57, 51 .

Overall mean $\bar{y}_{..} = 58.2$.

$CTSS = \sum\limits_{i=1}^{4} \sum\limits_{j=1}^{5} (y_{ij} - \bar{y}_{..})^2 = 1775.2$.

Sample means:

i	$\bar{y}_{i.}$	$\bar{y}_{i.} - \bar{y}_{..}$
1	60.6	2.4
2	51.4	-6.8
3	67.6	9.4
4	53.2	-5.0

$$MSS = 5 \sum_{i=1}^{4} (\overline{y}_i. - \overline{y}..)^2 = 826.8 .$$

Source of Variation	Degrees of Freedom	Sum of Squares	Mean Sum of Squares	f
Model	3	826.8	275.6	4.65
Residual	16	948.4	59.275	
Total	19	1775.2		

From Table 4, the critical value $F_{16}^{3}(0.05) = 3.24$.
So we would reject the null hypothesis, and accept that the model does explain the data; in other words, the sample means are not all the same, and significant differences do exist between the production lines.

10.3.1 NOTES ON CALCULATIONS

There are always at least two ways of computing any sum of squares (see the two formulae for S_{XX} etc. in chapter 9). For example,
$$CTSS = \sum_{i=1}^{k} \sum_{j=1}^{n} (y_{ij} - \overline{y}..)^2 = \sum_{i=1}^{k} \sum_{j=1}^{n} y_{ij}^2 - N\overline{y}_{..}^2 .$$
(You can confirm for yourself that these two formulae are equivalent).

A useful general principle for calculating $CTSS$ in *any* ANOVA problem is that
$$CTSS = \sum^{N} Data^2 - N \times (Overall\ mean)^2 .$$
Similarly, in this case, for the Model Sum of Squares:
$$MSS = n \sum_{i=1}^{k} (\overline{y}_i. - \overline{y}..)^2 = n \sum_{i=1}^{k} \overline{y}_i^2. - N\overline{y}_{..}^2 .$$

The term $N\overline{y}_{..}^2$ is sometimes called the *correction factor*, as it corrects both the total and model sums of squares for the overall mean $\overline{y}_{..}$.

The following program is designed to carry out an ANOVA test for differences between sample means, including the printing of the ANOVA table.

```pascal
program anova(input,output);
{ Program to test for differences between sample means. }
const maxsample = 10;
      maxvalues = 30;
type posint = 0..maxint;
     data = array[ 1..maxsample,1..maxvalues ] of real;
     samplearray = array[ 1..maxsample ] of real;
     valuearray = array[ 1..maxvalues ] of real;
var y : data;
    ydum : valuearray;
    ymean : samplearray;
    ybar,mss,ctss : real;
    i,j,k,n,ntot : posint;

function mean(var x : valuearray; n : posint) : real;
var i : posint;
    sum : real;
begin
  sum := 0;
  for i := 1 to n do sum := sum + x[ i ];
  mean := sum/n;
end;

function sumsq(var x : valuearray; xbar : real; n : posint)
                                          : real;
var i : posint;
    sum : real;
begin
  sum := 0;
  for i := 1 to n do sum := sum + sqr(x[ i ]-xbar);
  sumsq := sum;
end;

procedure anovatable(mss,ctss : real; ntot,k : posint);
var rss,f : real;
begin
  rss := ctss - mss; writeln;
  writeln('ANOVA table'); writeln;
  writeln('Source   Degrees of   Sum of    Mean Sum      f');
  writeln('         Freedom     Squares   of Squares');
  f := mss*(ntot-k)/(rss*(k-1)); writeln;
  writeln('Model',k-1:7,'        ',mss:10:3,mss/(k-1):10:3,
          f:10:3);
  writeln;
  writeln('Residual',ntot-k:4,'      ',rss:10:3,
          rss/(ntot-k):10:3);
  writeln('-------');
  writeln('Total',ntot-1:7,'     ',ctss:10:3);
end;

begin
  writeln; write('No. of samples ? '); read(k); writeln;
  write('No. in each sample ? '); read(n); writeln;
  ntot := n*k; ybar := 0; ctss := 0; mss := 0;
  for i := 1 to k do
  begin
    writeln; writeln('Input values for sample ',i:3);
    for j := 1 to n do
    begin
```

```pascal
          read(y[i,j]); ybar := ybar + y[i,j];
      end;
   end;
   ybar := ybar/ntot;
   writeln; writeln('Overall mean = ',ybar:10:4);
   for i := 1 to k do
   begin
      for j := 1 to n do ydum[j] := y[i,j];
      ymean[i] := mean(ydum,n);
      mss := mss + n*sqr(ymean[i]- ybar);
      ctss := ctss + sumsq(ydum,ybar,n);
      writeln('Mean for sample ',i:3,' = ',ymean[i]:10:4);
   end;
   anovatable(mss,ctss,ntot,k);
end.
```

10.4 Unequal Sample Sizes

It is a relatively simple matter to extend the preceding
analysis to the case where the different samples have different
numbers of values in them. Suppose that the ith sample
contains n_i values.

$$\text{Then } \quad N = \sum_{i=1}^{k} n_i , \qquad CTSS = \sum_{i=1}^{k} \sum_{j=1}^{n_i} (y_{ij} - \overline{y}_{..})^2$$

$$= \sum_{i=1}^{k} \sum_{j=1}^{n_i} y_{ij}^2 - N\overline{y}_{..}^2 .$$

$$\text{Sample means:} \quad \overline{y}_{i.} = \frac{1}{n_i} \sum_{j=1}^{n_i} y_{ij} ,$$

$$\text{and} \quad MSS = \sum_{i=1}^{k} n_i (\overline{y}_{i.} - \overline{y}_{..})^2 = \sum_{i=1}^{k} n_i \overline{y}_{i.}^2 - N\overline{y}_{..}^2 .$$

Example: A number of similar programs have been written in
different languages, and the total man-hours taken
to develop each have been measured, as follows:

FORTRAN: 65, 72, 49, 75, 59

PASCAL: 38, 51, 48, 63

COBOL: 81, 71, 67, 73, 62, 55, 81 .

Is there evidence to support the idea that some
languages take less time than others for program
development ?

$N = 16; \quad \overline{y}_{..} = 63.125 .$

i	n_i	$\bar{y}_{i.}$	$\bar{y}_{i.} - \bar{y}_{..}$
1	5	64.0	0.875
2	4	50.0	-13.125
3	7	70.0	6.875

$$MSS = 5 \times 0.875^2 + 4 \times (-13.125)^2 + 7 \times 6.875^2 = 1023.75 .$$

$$CTSS = \sum_i \sum_j y_{ij}^2 - N\bar{y}_{..}^2 = 66084 - 63756.25 = 2327.75 .$$

Source of Variation	Degrees of Freedom	Sum of Squares	Mean Sum of Squares	f
Model	2	1023.75	511.875	⎤ 5.10
Residual	13	1304.0	100.308	⎦
Total	15	2327.75		

Now the critical $F_{13}^2(0.05) = 3.81$, and we reject H_0, the hypothesis that all the samples have the same mean. So there is a significant difference due to the different languages, at the 5% level of significance.

10.5 Two-way ANOVA

Quite often experiments whose results we wish to analyse involve the effects of two different *factors*, or sources of variation in the data, apart from the residual error. The classical example of this comes from agriculture. Suppose that a farmer wishes to test out some different treatments (e.g. fertilisers) on his crops; he may plant the same crop in several different fields and apply a different treatment to each field, as in the picture:

Figure 10.1

Field 1	Field 2	Field 3	Field 4
T_1	T_2	T_3	T_4

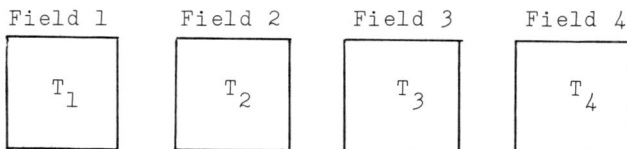

If he takes n measurements of crop yield from each field, this experiment could be analysed as above, testing for significant differences between fields. However, there

155

are problems with this approach:

a) What does he do if he only has 3 fields and wishes to test 4 treatments ?

b) More seriously, if he gets a significant difference between the fields he does not know if this is due to the different treatments, or if it is just that some fields are naturally more fertile than others. Technically, we say that the effects of the treatments are *confounded* with the effects of the fields.

A better way of doing this experiment, using only 3 fields, is:

Figure 10.2

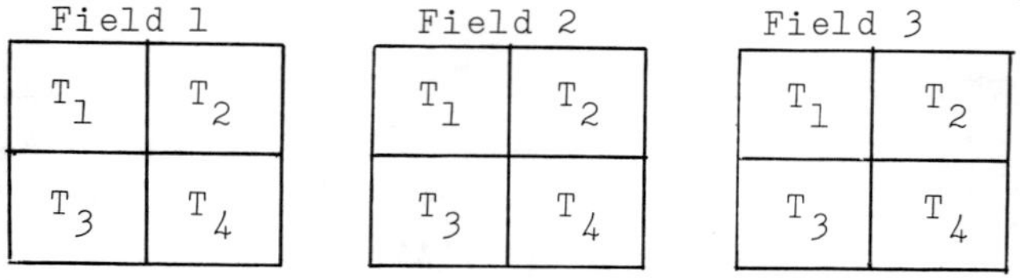

Field 1		Field 2		Field 3	
T_1	T_2	T_1	T_2	T_1	T_2
T_3	T_4	T_3	T_4	T_3	T_4

This is called a *balanced design*, because there are the same numbers of each treatment in each field, or *block*. An unbalanced design might look like:

Figure 10.3

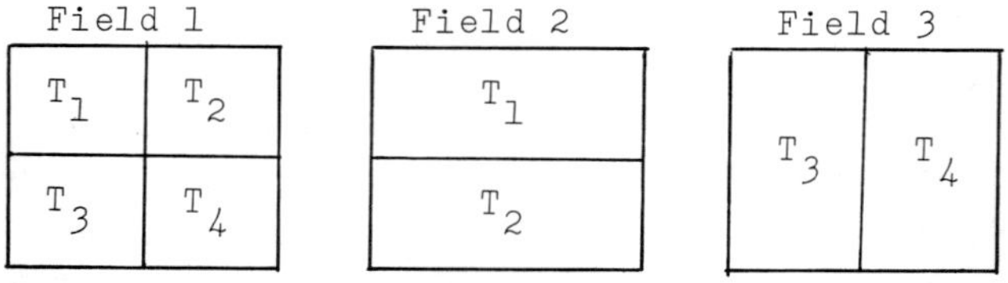

Field 1		Field 2	Field 3	
T_1	T_2	T_1		
T_3	T_4	T_2	T_3	T_4

Within a balanced design we may repeat, or *replicate*, treatments within blocks:

Figure 10.4

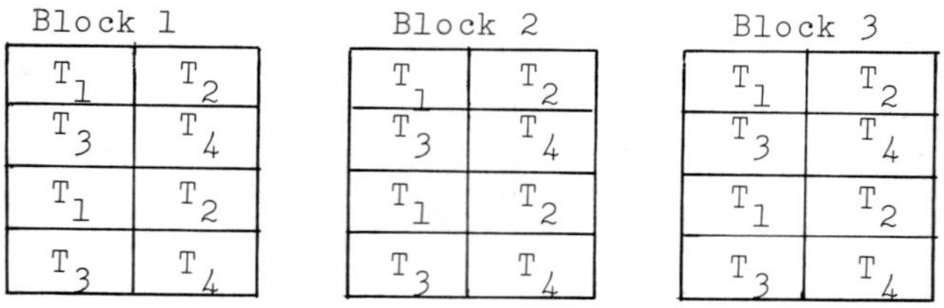

Block 1		Block 2		Block 3	
T_1	T_2	T_1	T_2	T_1	T_2
T_3	T_4	T_3	T_4	T_3	T_4
T_1	T_2	T_1	T_2	T_1	T_2
T_3	T_4	T_3	T_4	T_3	T_4

There are two factors to be considered in this experiment: treatments (fertilisers) and blocks (fields). However, although these two terms, "treatments" and "blocks", are used conventionally, the analysis of the experiment is the same for both. Normally, we call the factor of major interest "treatments", and the other "blocks". Given the results of a balanced experiment, we can answer the following questions:

1. Is there a significant difference due to the different treatments ?
2. Is there a significant difference due to the different blocks ?

These two questions can be answered using a single ANOVA table from the results of a balanced experiment.

Suppose y_{ij} is the result, or *yield*, from treatment i applied to block j, and that there are m treatments and n blocks. So $N = mn$. Our simple model is:

$$y_{ij} = \mu + T_i + B_j + \varepsilon_{ij} \ ,$$

where μ is the overall mean,

$\quad T_i$ is the effect of the ith treatment,

$\quad B_j$ is the effect of the jth block,

and $\quad \varepsilon_{ij}$ is the residual error.

Assume that $\sum\limits_{i=1}^{m} T_i = 0$ and $\sum\limits_{j}^{n} B_j = 0$.

We shall estimate the model parameters as follows:

Estimate μ by $\overline{y}_{..} = \dfrac{1}{N} \sum\limits_{i=1}^{m} \sum\limits_{j=1}^{n} y_{ij}$.

Estimate T_i by $\overline{y}_{i.} - \overline{y}_{..}$ $\quad (\overline{y}_{i.} = \dfrac{1}{n} \sum\limits_{j=1}^{n} y_{ij})$.

Estimate B_j by $\overline{y}_{.j} - \overline{y}_{..}$ $\quad (\overline{y}_{.j} = \dfrac{1}{m} \sum\limits_{i=1}^{m} y_{ij})$.

Sums of squares:

$$CTSS = \sum\limits_{i=1}^{m} \sum\limits_{j=1}^{n} (y_{ij} - \overline{y}_{..})^2 = \sum\limits_{i=1}^{m} \sum\limits_{j=1}^{n} y_{ij}^2 - N\overline{y}_{..}^2 \quad \text{(as usual)} .$$

Treatments: $TSS = n \sum\limits_{i=1}^{m} (\overline{y}_{i.} - \overline{y}_{..})^2 = n \sum\limits_{i=1}^{m} \overline{y}_{i.}^2 - N\overline{y}_{..}^2$.

Blocks: $\quad BSS = m \sum\limits_{j=1}^{n} (\overline{y}_{.j} - \overline{y}_{..})^2 = m \sum\limits_{j=1}^{n} \overline{y}_{.j}^2 - N\overline{y}_{..}^2$.

For both treatment and block sums of squares, the rule to remember is simple: multiply outside the summation by the number over which you have averaged to get the values inside the summation. Thus to get TSS, we multiply by n, as each treatment is averaged over the n blocks.

Therefore we no longer have a single model sum of squares, but two separate sums corresponding to treatments and blocks. The relationship is:

$$CTSS = TSS + BSS + RSS .$$

Degrees of freedom:
$$\text{Total} = N-1 = mn-1$$
$$\text{Treatments} = m-1$$
$$\text{Blocks} = n-1$$
$$\text{Residual} = mn-1 - (m-1) - (n-1)$$
$$= (m-1)(n-1) .$$

The ANOVA table will provide us with two f-ratios, so that we can test simultaneously the hypothesis that the treatment effects are all zero and the hypothesis that the block effects are all zero.

Example: 4 treatments on 3 blocks: yield y_{ij} = tons/acre of turnips from ith treatment on jth block.

Results:

		Blocks		
		1	2	3
	1:	38	52	45
Treatments	2:	61	73	64
	3:	43	51	59
	4:	30	40	44

Overall average $\overline{y}_{..} = 600/12 = 50.0$.

$CTSS$ $\quad \sum\sum y_{ij}^2 - N\overline{y}_{..}^2 = 31686 - 12\times50^2 = 1686.0$.

Treatments:

i	$\overline{y}_{i.}$	$\overline{y}_{i.}-\overline{y}_{..}$
1	45.0	-5.0
2	66.0	16.0
3	51.0	1.0
4	38.0	-12.0

$$TSS = 3\sum_{i=1}^{4}(\overline{y}_{i.}-\overline{y}_{..})^2 = 1278.0 .$$

Blocks:

j	$\overline{y}_{.j}$	$\overline{y}_{.j}-\overline{y}_{..}$
1	43.0	-7.0
2	54.0	4.0
3	53.0	3.0

$$BSS = 4\sum_{j=1}^{3}(\overline{y}_{.j}-\overline{y}_{..})^2 = 296.0 .$$

Source of Variation	Degrees of Freedom	Sum of Squares	Mean Sum of Squares	f
Treatments	3	1278.0	426.0	22.82
Blocks	2	296.0	148.0	7.93
Residual	6	112.0	18.667	
Total	11	1686.0		

(The fourth column, mean sum of squares, is always obtained by dividing the sum of squares by the degrees of freedom).

a) To test the hypothesis that all the treatment effects = 0, the critical value $F_6^3(0.05) = 4.75$. Since the calculated f value for treatments (22.82) is greater than this, we reject H_0 and say that the treatments definitely affect the yield of the experiment.

b) To test the hypothesis that all the block effects = 0, the critical value $F_6^2(0.05) = 5.14$. Again the calculated f value for the blocks is greater than this, and we conclude that there is a significant effect due to the blocks.

10.6 Analysis with Replications

The balanced experiment may be replicated r times - this means that each combination of blocks and treatments has r separate measurements made on it.
Total number of measurements $N = m \times n \times r$.

If $y_{ij\ell}$ is the result of the ℓth replicate of the ith treatment on the jth block, then the overall average

$$\overline{y}_{...} = \frac{1}{N} \sum_{i=1}^{m} \sum_{j=1}^{n} \sum_{\ell=1}^{r} y_{ij} \quad .$$

$$CTSS = \sum_{i=1}^{m} \sum_{j=1}^{n} \sum_{\ell=1}^{r} (y_{ij\ell} - \overline{y}_{...})^2 = \sum_{i=1}^{m} \sum_{j=1}^{n} \sum_{\ell=1}^{r} y_{ij\ell}^2 - N\overline{y}_{...}^2 \quad .$$

Treatments: $\quad \overline{y}_{i..} = \frac{1}{nr} \sum_{j=1}^{n} \sum_{\ell=1}^{r} y_{ij\ell} \quad .$

$$TSS = nr \sum_{i=1}^{m} (\overline{y}_{i..} - \overline{y}_{...})^2 = nr \sum_{i=1}^{m} \overline{y}_{i..}^2 - N\overline{y}_{...}^2 \quad .$$

(with $m-1$ degrees of freedom, as before).

159

Blocks: $\quad \overline{y}_{.j.} \;=\; \dfrac{1}{mr} \sum\limits_{i=1}^{m} \sum\limits_{\ell=1}^{r} y_{ij\ell}$.

$BSS \;=\; mr \sum\limits_{j=1}^{n} (\overline{y}_{.j.} - \overline{y}_{...})^2 \;=\; mr \sum\limits_{j=1}^{n} \overline{y}^2_{.j.} \;-\; N\overline{y}^2_{...}$

(with $n-1$ degrees of freedom, as before).

Example: 3 languages (FORTRAN, Pascal, COBOL) are being
tested for speed of development using 5 programmers
(A,B,C,D and E). Each programmer writes 2 programs
in each language.
Treatments = languages ($m = 3$).
Blocks = programmers ($n = 5$).
Replicates = programs ($r = 2$).
    Results (man-hours/program written):

Programmers

	A	B	C	D	E
FORTRAN:	25,31	18,15	34,30	43,33	23,18
Pascal:	16,24	14,20	23,29	31,25	14,18
COBOL:	29,25	33,20	29,35	37,35	21,26

$N \;=\; 3\times5\times2 \;=\; 30; \quad \overline{y}_{...} \;=\; 774/30 \;=\; 25.8$.

$CTSS \;=\; \underline{21699} - 30\times25.8^2 \;=\; 1729.8$.

Languages (Treatments):

i	$\overline{y}_{i..}$	$\overline{y}_{i..} - \overline{y}_{...}$
1	27.0	1.2
2	21.4	-4.4
3	29.0	3.2

$TSS \;=\; 10 \sum\limits_{i=1}^{3} (\overline{y}_{i..} - \overline{y}_{...})^2 \;=\; 310.4$.

Programmers (Blocks):

j	$\overline{y}_{.j.}$	$\overline{y}_{.j.} - \overline{y}_{...}$
1	25.0	-0.8
2	20.0	-5.8
3	30.0	4.2
4	34.0	8.2
5	20.0	-5.8

$BSS \;=\; 6 \sum\limits_{j=1}^{5} (\overline{y}_{.j.} - \overline{y}_{...})^2 \;=\; 916.8$.

Source of Variation	Degrees of Freedom	Sum of Squares	Mean Sum of Squares	f
Languages	2	310.4	155.2	7.10
Programmers	4	916.8	229.2	10.49
Residual	23	502.6	21.85	
Total	29	1729.8		

a) To test languages, use critical $F_{23}^{3}(0.05) = 3.42$. So there is a significant effect due to different languages.

b) To test programmers, use critical $F_{23}^{4}(0.05) = 2.80$. So there is a significant effect due to programmers.

10.7 Exercises

1. 3 different techniques are being tested for chip manufacture at 4 different sites. The percentage of usable chips is measured for each technique, with the following results:

		Sites			
		A	B	C	D
	1:	12	18	24	14
Techniques	2:	20	25	33	22
	3:	7	11	21	9

$(\sum \text{Data} = 216, \sum \text{Data}^2 = 4530)$.

Analyse the results of this experiment in two different ways, using a significance level of 1%:
a) Using simple ANOVA for differences in techniques only, ignoring the differences between sites.
b) Using two-way ANOVA, testing for differences between sites as well as techniques.
Comment on the different results obtained.

2. Do Exercise 3, chapter 7, using ANOVA. How does the f-statistic relate to the t-statistic obtained previously?

3. Do Exercise 5, chapter 7, using two-way ANOVA.

4. 20 new cars in a showroom have been thoroughly inspected, and a quantity "percentage defective workmanship" has been evaluated for each. The cars have been classified according to the day of the week on which they were made, and the results tabulated as follows:

   Monday: 15, 24, 13, 18
   Tuesday: 10, 8, 12
   Wednesday: 13, 11, 9, 7
   Thursday: 15, 16, 23, 19, 12
   Friday: 22, 28, 31, 24
   ($\sum$Data = 330, $\sum$Data^2 = 6342).

   Is there any effect due to the day of manufacture ?

5. 3 processes for manufacturing an alloy have been tested at 5 different plants and the percentage impurity has been measured in each case. The experiment is replicated twice, with the following results:

		Plants			
	1	2	3	4	5
1:	8,12	11,7	5,9	7,7	8,6
Processes 2:	6,8	9,3	5,3	3,5	6,2
3:	11,9	8,10	10,4	8,6	9,5

   ($\sum$Data = 210, $\sum$Data^2 = 1668).

   Analyse this data as fully as possible.

10.8 Computer Projects

1. Amend the ANOVA program in the text to cope with different numbers in each sample.

2. Write a program to carry out two-way ANOVA with replications, printing out a suitable ANOVA table.

3. Design a general ANOVA program to carry out all types of ANOVA, with a flexible user interface.

Suggested Further Reading for Section II

"Modern Elementary Statistics", by John E. Freund,
     Prentice-Hall, 1974.

"General Statistics", by Audrey Haber and Richard P. Runyon,
     Addison-Wesley, 1977.

"How to Lie with Statistics", by Darrell Huff, Penguin Books,
     1973.

"Statistical Methods", by Allen L. Edwards, Holt, Rinehart
     and Winston, 1967.

"Fundamental Statistical Concepts", by Frederic E. Fischer,
     Canfield Press, 1973.

     (And many, many more textbooks on Statistics)

SECTION III

Stochastic Processes

CHAPTER 11

The Simple Random Walk

11.1 Introduction

The study of stochastic processes is the study of models of systems which vary in time, in a random or stochastic fashion. Such models can be usefully applied to all kinds of problems, including analysing the behaviour of computer systems. Some of the concepts and methods used are well illustrated by looking at the simple random walk, which despite its name is not always simple in its analysis.

The behaviour of the simple random walk, or "Drunkard's walk", can be described as follows: consider an object (drunkard) initially at position 0 (a lamp-post) at time 0. During each increment of time, it takes a unit step, either to the right with probability p, or to the left with probability q $(= 1-p)$.

Figure 11.1

Each step is independent of all the others. If we plot a graph of distance from the origin versus time, we might get a

167

picture like this:

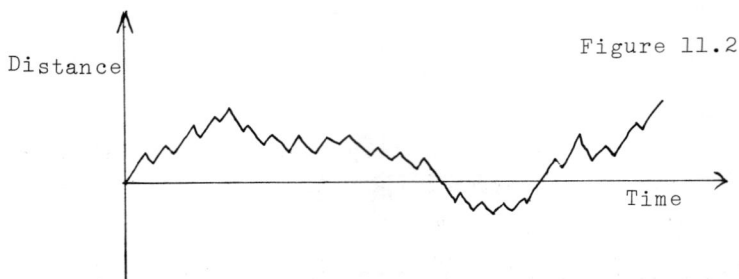

Figure 11.2

11.2 Probabilities of Return to the Origin

When studying the simple random walk, one of the things we can compute is the probability of returning to the origin at various points in time. It is obvious that if the object is going to return to 0, it must do so in an even number of steps, since the number of positive steps taken must equal the number of negative steps taken. So the probability of returning to 0 in an odd number of steps is zero.

Let u_{2n} = P(Object returns to 0 after $2n$ steps). For this to happen, we need to have n positive steps and n negative steps, arranged in any order. Thus we have a Binomial probability, with $2n$ trials and n "successes" (positive steps), each with probability p.

So $u_{2n} = {}^{2n}C_n \, p^n q^n$.

This is fine, but unfortunately the set of probabilities u_2, u_4, u_6, ... does not form a probability distribution, and the sum

$$\sum_{n=1}^{\infty} u_{2n} \neq 1 .$$

The reason for this is that the events represented by these probabilities are not mutually exclusive - for example, it is possible to return to 0 after 2 steps and again after 4 steps, etc.

168

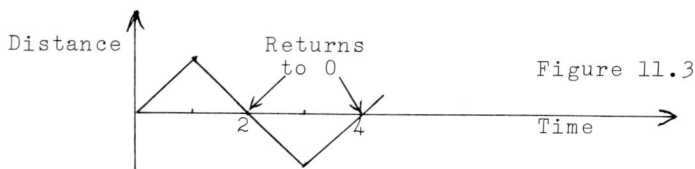

Figure 11.3

We can define a set of mutually exclusive events which do give rise to a probability distribution, as follows:

f_{2n} = P(Returns to 0 after 2n steps, *for the first time*) .

If we define f_∞ = P(Never returns to 0), then we have a complete probability distribution and

$$\sum_{n=1}^{\infty} f_{2n} = 1 .$$

Now, it can be shown (though I do not intend to prove it) that

$$f_\infty = \sqrt{1-4pq} .$$

Therefore, if $p = q = \frac{1}{2}$, $f_\infty = 0$, and this implies that the object will ultimately return to 0 at some time (although there is no telling how long it may take to do so). On the other hand, if $p \neq q$, then $f_\infty > 0$, and there is a non-zero probability that the object will never return to 0 at all.

It would now seem nice to produce a formula for the probability f_{2n} of first return to the origin after 2n steps, but this is by no means as simple a task as that of deriving the formula for u_{2n}. To compute f_{2n}, we need to consider the relationship between this probability and u_{2n}.

Now u_{2n} = P(Object at 0 after 2n steps, whether for the first time or not) .

We now *decompose* u_{2n} with respect to the first return time. This means that we consider the probability:

P(Object at 0 after 2n steps, and first return to 0 was after 2t steps), where $t \leq n$.

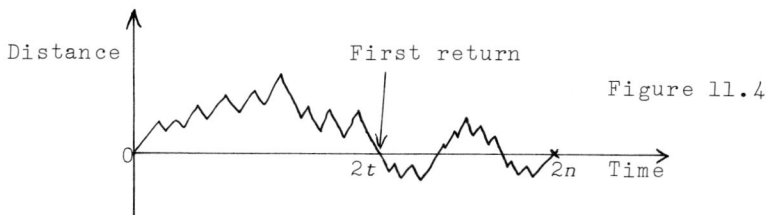

Figure 11.4

This probability is

$$f_{2t} \times P(\text{Object returns to 0 again after } 2n\text{-}2t \text{ steps})$$
$$= f_{2t} u_{2(n-t)} \quad .$$

We can reconstruct u_{2n} by summing these probabilities for all values of t, getting

$$u_{2n} = \sum_{t=1}^{n} f_{2t} u_{2(n-t)} \quad ,$$

using the conventions: $u_0 = 1$, $f_0 = 0$. We can rewrite this relationship to get values of f_{2n} systematically:

$$f_{2n} = u_{2n} - \sum_{t=1}^{n-1} f_{2t} u_{2(n-t)} \quad .$$

<u>$n = 1$</u>

$$u_2 = {}^2C_1 \, p^1 q^1 = 2pq \quad .$$
$$f_2 = u_2 = 2pq \quad .$$

<u>$n = 2$</u>

$$u_4 = {}^4C_2 \, p^2 q^2 = 6p^2 q^2 \quad .$$
$$f_4 = u_4 - f_2 u_2 = 6p^2 q^2 - 2pq \times 2pq = 2p^2 q^2 \quad .$$

Continuing in this way, we can tabulate values of u_{2n} and f_{2n} for the first few values of n:

$2n$	u_{2n}	f_{2n}
2	$2pq$	$2pq$
4	$6p^2 q^2$	$2p^2 q^2$
6	$20p^3 q^3$	$4p^3 q^3$
8	$70p^4 q^4$	$10p^4 q^4$

etc.

We now have a method for computing f_{2n}, but it is clearly going to be tedious to use for large n, since it will involve computing all the preceding values as well. It would be nicer to have a more direct formula for f_{2n}. Examining the results obtained above, it is possible to calculate the ratio u_{2n}/f_{2n} for the first few values of n:

$2n$	u_{2n}/f_{2n}
2	1
4	3
6	5
8	7

It seems from this that $u_{2n}/f_{2n} = 2n-1$, which leads to the

170

hypothetical formula:
$$f_{2n} = u_{2n}/(2n-1) \ .$$
To show that this formula works for all values of n is going to take us through a long and circuitous chain of reasoning, but the process of doing so will illustrate a lot of features of the simple random walk.

11.3 Equal Probabilities and the Reflection Principle

For this section and most of the next, we shall concentrate on the special case of the simple random walk with equal probabilities: that is, $p = q = \frac{1}{2}$. In this special case, we can make use of what is known as the reflection principle.

Suppose we are interested in computing the probability of the random walk going from point A (a units from the origin, time 0) to point B (b units from the origin, time $2n$) in $2n$ steps, *without returning to 0 in between*.

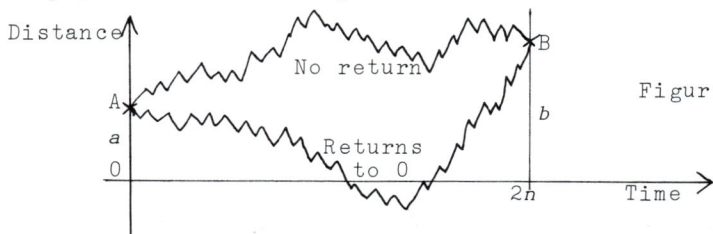

Figure 11.5

Now, if we go from A to B we can either do so without returning to 0 in between, or we can do so with an intermediate return to 0. Therefore,

P(Going from A to B) = P(A to B with no return to 0)
                       + P(A to B with a return to 0).

Let us therefore consider the latter event, of going from A to B with~~out~~ an intermediate return to 0. We may investigate an example of such a path, including a return to 0 at the point C.

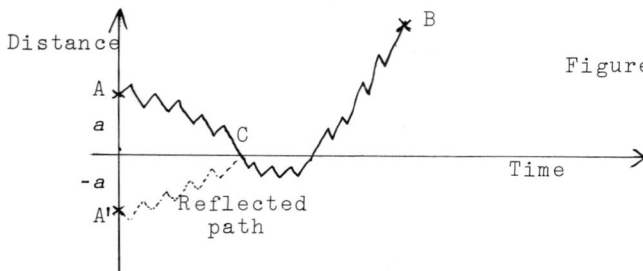

Figure 11.6

Now reflect the part of the path from A to C in the axis, so that each negative step becomes positive and vice versa. This leads to a path from the reflected point A' (*a* units below the axis) to B. Each path from A to B which returns to O in between has a unique reflected path from A' to B, and each path from A' to B is equivalent in the same way to a path from A to B with an intermediate return to O. Since $p = q = \frac{1}{2}$, the probability of each path is the same, and so it becomes clear that P(Going from A to B with a return to O)

= P(Going from A' to B).

So finally we can say:
P(Going from A to B *without* a return to O)
= P(going from A to B) - P(going from A' to B).

This is derived from the reflection principle, and in the next section we shall go on to use it in deriving further results. Note that it is only valid for equal probabilities, with $p = q = \frac{1}{2}$.

11.4 Probabilities of Not Returning to Zero

Again, for most of this section we shall assume the special case: $p = q = \frac{1}{2}$, so that we may use the results of the reflection principle. Define:

S_{2n} = Position of the object after $2n$ steps
= Number of steps above O.

Consider the probability:

v_{2n} = P(Never returns to O in the first $2n$ steps)

172

$$= P(S_2 \neq 0, S_4 \neq 0, \dots S_{2n} \neq 0).$$

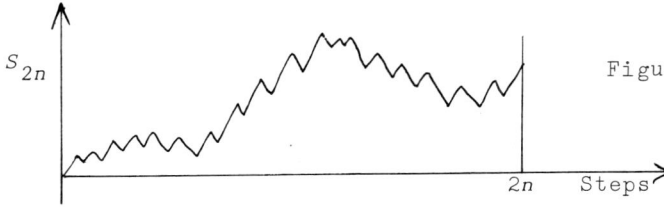

Figure 11.7

Clearly, if this is to be true, the path must either be all positive or all negative for the first $2n$ steps - there can be no crossing over.

So $\qquad v_{2n} = v_{2n}^+ + v_{2n}^-$,

where $v_{2n}^+ = P(S_2 > 0, S_4 > 0, \dots S_{2n} > 0)$

and $\qquad v_{2n}^- = P(S_2 < 0, S_4 < 0, \dots S_{2n} < 0)$.

Consider v_{2n}^+, and decompose this with respect to the position after $2n$ steps, $2r$ say. That is, find

$$P(S_2 > 0, S_4 > 0, \dots S_{2n-2} > 0, S_{2n} = 2r), \text{ for } 0 < r \leq n.$$

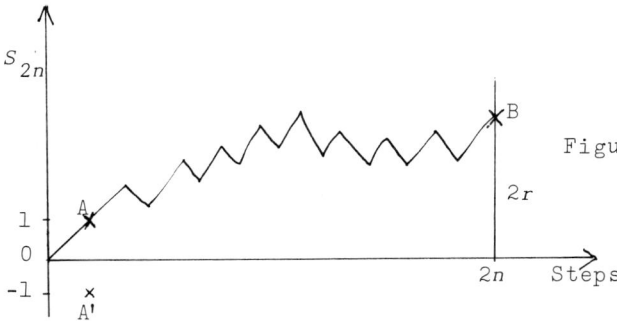

Figure 11.8

For this to happen, we need:

a) $S_1 = +1$ - probability $= \frac{1}{2}$.

b) The object goes from A (i unit above the origin at time 1) to B ($2r$ units above the origin at time $2n$) in $2n-1$ steps without returning to 0.

To find the probability of b) we need the reflected point A', at position -1 after 1 step.

$$P(A \text{ to } B \text{ with no return to } 0) = P(A \text{ to } B) - P(A' \text{ to } B).$$

173

Now P(A to B) = P(1 to 2r in 2n-1 steps)
                = P(0 to 2r-1 in 2n-1 steps)
                = $P(S_{2n-1} = 2r-1)$.

And P(A' to B) = P(-1 to 2r in 2n-1 steps)
                = P(0 to 2r+1 in 2n-1 steps)
                = $P(S_{2n-1} = 2r+1)$.

So $P(S_2 > 0, S_4 > 0, \ldots S_{2n} = 2r)$
                = $\tfrac{1}{2}(P(S_{2n-1} = 2r-1) - P(S_{2n-1} = 2r+1))$.

Now v^{+}_{2n} = $\sum_{r=1}^{n} P(S_2 > 0, S_4 > 0, \ldots S_{2n} = 2r)$

                = $\tfrac{1}{2}(P(S_{2n-1} = 1) - \cancel{P(S_{2n-1} = 3)}$
                + $\tfrac{1}{2}(\cancel{P(S_{2n-1} = 3)} - \cancel{P(S_{2n-1} = 5)}$
                        $\vdots$ $\vdots$
                + $\tfrac{1}{2}(\cancel{P(S_{2n-1} = 2n-1)} - P(S_{2n-1} = 2n+1))$.

Most of the terms cancel as shown; the last term is 0 because it is impossible to get 2n+1 steps away from the origin in just 2n-1 steps. So we are left with:

$$v^{+}_{2n} = \tfrac{1}{2}P(S_{2n-1} = 1) .$$

An exactly similar argument gets us the result that

$$v^{-}_{2n} = \tfrac{1}{2}P(S_{2n-1} = -1) .$$

So v_{2n} = $\tfrac{1}{2}P(S_{2n-1} = 1) + \tfrac{1}{2}P(S_{2n-1} = -1)$.

Consider the following diagram:

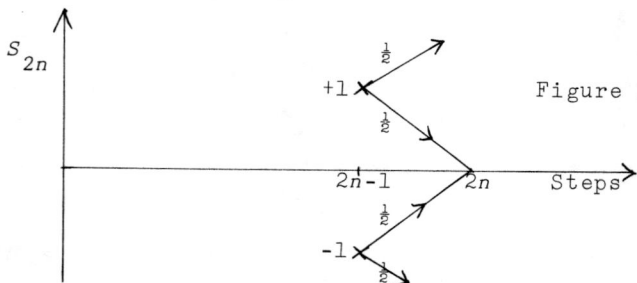

Figure 11.8

The probability of reaching 0 after 2n steps from above is equal to the probability that $S_{2n-1} = 1$, times the probability of a negative step ($\tfrac{1}{2}$). Similarly, the probability of reaching 0 after 2n steps from below is equal to $\tfrac{1}{2}P(S_{2n-1} = -1)$.

Thus u_{2n} $=$ $P(S_{2n} = 0)$ $=$ $\frac{1}{2}P(S_{2n-1} = 1) + \frac{1}{2}P(S_{2n-1} = -1)$
$=$ v_{2n} .

So we end up with the rather strange result that
P(Object never returns to 0 in the first $2n$ steps)
$=$ P(Object is at 0 after $2n$ steps) .

11.5 Formulae for f_{2n} and u_{2n}

The interesting result derived in the last section will now lead us back to a formula for f_{2n}. To do this, we consider the probability of never returning to 0 in the first $2n-2$ steps:
$$P(S_2 \neq 0, S_4 \neq 0, \ldots S_{2n-2} \neq 0) = v_{2n-2} = u_{2n-2} .$$
Any path satisfying this may either continue to stay away from 0 for the next 2 steps, or may return to 0 for the first time after $2n$ steps.

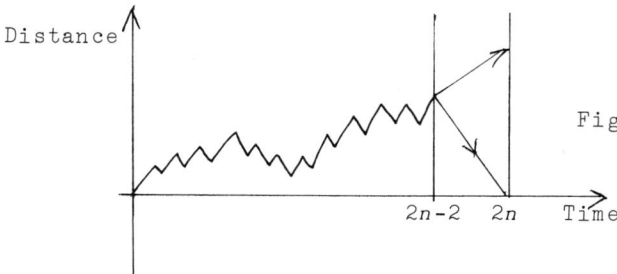

Figure 11.10

Thus $\quad u_{2n-2}$ $=$ $u_{2n} + f_{2n}$

$\Rightarrow \quad f_{2n}$ $=$ $u_{2n-2} - u_{2n}$.

Now $\quad u_{2n-2}$ $=$ $\dfrac{(2n-2)!}{(n-1)!(n-1)!}$ $(\frac{1}{2})^{2n-2}$

and $\quad u_{2n}$ $=$ $\dfrac{2n!}{n! \, n!}$ $(\frac{1}{2})^{2n}$.

The ratio u_{2n-2}/u_{2n} $=$ $\dfrac{4n^2}{2n(2n-1)}$ $=$ $\dfrac{2n}{2n-1}$.

Therefore, $\quad u_{2n-2}$ $=$ $\dfrac{2n}{2n-1} u_{2n}$,

and $\quad f_{2n}$ $=$ $u_{2n-2} - u_{2n}$

175

$$= \frac{2n}{2n-1} u_{2n} - u_{2n} = \frac{u_{2n}}{2n-1} \; .$$

Thus we have finally derived the formula for f_{2n} that we hypothesised right at the start. However, it has only been derived for the special case $p = q = \frac{1}{2}$. How is it therefore that it seems to work for all values of p and q ?

The way to see this is to consider the paths that get us from 0 back to 0 again in $2n$ steps, with or without an intermediate return to 0. For example, with $n = 3$, we have the following paths:

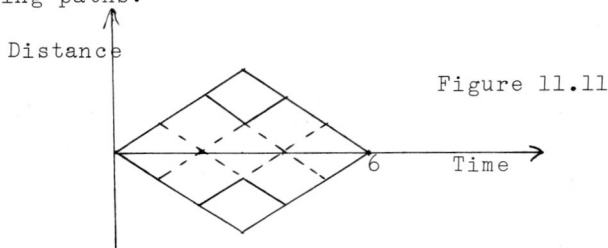

Figure 11.11

Paths not returning to 0 in between are solid, and others are dashed.

If N_f is the number of paths from 0 to 0 in $2n$ steps which do not return to 0 in between, and N_u is the total number of paths from 0 to 0, then
$$f_{2n} = N_f p^n q^n \text{ and } u_{2n} = N_u p^n q^n \; ,$$
since each path has the same probability $p^n q^n$.

So the ratio $u_{2n}/f_{2n} = N_u/N_f$, which is independent of p and q. Since we have shown that this ratio is equal to $2n-1$ when $p = q = \frac{1}{2}$, it follows that it must be the same for all values of p and q.

Therefore, $f_{2n} = \dfrac{{}^{2n}C_n p^n q^n}{2n-1} = \dfrac{u_{2n}}{2n-1} \; .$

Before leaving this section, it is useful to derive an approximate formula for u_{2n}, again for the special case $p = q = \frac{1}{2}$.
$$u_{2n} = {}^{2n}C_n (\tfrac{1}{2})^{2n} = \frac{2n!}{n!\,n!} (\tfrac{1}{2})^{2n} \; .$$

For large values of n, Stirling's approximation gives an approximate value of $n!$ as:

$$n! \simeq \sqrt{2\pi}\; n^{n+\frac{1}{2}}\; e^{-n}\;.$$

Applying this to the formula for u_{2n}, we get

$$u_{2n} \simeq \frac{\sqrt{2\pi}\;(2n)^{2n+\frac{1}{2}}\;e^{-2n}}{\sqrt{2\pi}\;n^{n+\frac{1}{2}}\;e^{-n}\;\sqrt{2\pi}\;n^{n+\frac{1}{2}}\;e^{-n}}\;(\tfrac{1}{2})^{2n} = \frac{1}{\sqrt{\pi n}}\;.$$

We can check out this approximate formula for various values of n.

$2n$	u_{2n} (Exact)	u_{2n} (Approximate)
4	0.3750	0.3989
10	0.2461	0.2535
20	0.1762	0.1784

11.6 Simulation of a Simple Random Walk

As a way of checking the mathematics we have done, we can *simulate* the behaviour of the simple random walk. This involves writing a computer program which will generate a large number of random walks, starting from 0 each time. If we generate say N such random walks, we can approximately compute the probabilities.

For example,

$u_{2n} \simeq$ (Number of times walk at 0 after $2n$ steps)$/N$,

$f_{2n} \simeq$ (Number of times at 0 for the first time)$/N$,

etc.

To do this we need to be able to generate so-called *pseudo-random numbers* in the interval $[0,1]$. These numbers should behave like true random numbers uniformly and independently distributed, but each should be derived explicitly from the one before, so that the whole sequence is repeatable, if necessary. The generation of such sequences of pseudo-random numbers with reasonable properties is quite an important topic in its own right, which we shall not cover here. A simple way of doing it is as follows:

Let a and b be two fixed numbers, such that they do not exactly divide each other (e.g. $a = 519.7$, $b = 37.0$). Let x

be a number between 0 and *b*. Then let *z* be the fractional
part of *ax/b*, and define the next *x* to be equal to *bz*. We
shall assume that the sequence of *z* values so formed gives us
a suitable set of pseudo-random numbers in the range [0,1],
defined by the initial value of *x*, known as the *seed*.

The following program makes use of this to simulate the
behaviour of a simple random walk.

```
program walk(input,output);
{ Program to simulate a simple random walk. }
const mtab = 50;
      a = 519.7;
      b = 37.0;
type steparray = array[1..mtab] of real;
     posint = 0..maxint;
var x,p : real;
    nsteps,nits,k : posint;
    f,u,v : steparray;

function random(var x : real) : real;
{ Function to generate uniform pseudo-random numbers. }
var z : real;
    i : posint;
begin
  z := a*x/b;
  i := trunc(z);
  z := z - i;
  x := z*b;
  random := z;
end;

function step(var x : real) : integer;
{ Function to generate a random walk step. }
begin
  if random(x) > p then step := 1
                   else step := -1;
end;

procedure iterate(var x : real; nsteps : posint);
{ Procedure to generate one realisation of a random walk. }
var pos : integer;
    i : posint;
    imback : boolean;
begin
  pos := 0; i := 0; imback := false;
  repeat
    pos := pos + step(x) + step(x);
    i := i + 1;
    if pos = 0 then
    begin
      u[i] := u[i] + 1;
      if not imback then f[i] := f[i] + 1;
      imback := true;
    end
```

178

```
        else
          if not imback then v[i] := v[i] + 1;
    until 2*i >= nsteps;
end;

begin
  writeln('No. of steps to be simulated ? ');
  read(nsteps); writeln(nsteps:5);
  writeln('No. of iterations ? ');
  read(nits); writeln(nits:5);
  writeln('Probability of a positive step ? ');
  read(p); writeln(p:10:5);
  writeln('Initial seed ? ');
  read(x); writeln(x:10:5);
  for k := 1 to mtab do
  begin
    u[k] := 0; f[k] := 0; v[k] := 0;
  end;
  for k := 1 to nits do
    iterate(x,nsteps);
  writeln; writeln('   2n      u           f           v ');
  for k := 1 to nsteps div 2 do
  begin
    u[k] := u[k]/nits; f[k] := f[k]/nits;
    v[k] := v[k]/nits;
    writeln(2*k:5,u[k]:10:5,f[k]:10:5,v[k]:10:5);
  end;
end.
```

The results from a run of this program with 200
iterations were as follows:

```
No. of steps to be simulated ?
   20
No. of iterations ?
  200
Probability of a positive step ?
  0.50000
Initial seed ?
  12.45100

    2n       u           f           v
     2    0.43500     0.43500     0.56500
     4    0.38000     0.13000     0.43500
     6    0.37500     0.09500     0.34000
     8    0.27000     0.03000     0.31000
    10    0.21000     0.03000     0.28000
    12    0.19500     0.02500     0.25500
    14    0.17500     0.01500     0.24000
    16    0.14000     0.01000     0.23000
    18    0.14500     0.01000     0.22000
    20    0.20000     0.02500     0.19500
```

We would not expect these results to compare exactly
with those obtained mathematically, but the agreement is
reasonably good.

11.7 Exercises

1. 12 voters arrive at random to vote for two candidates, Fred and Bert. Each voter has a 50% chance of voting for each candidate. What is the probability that Fred wins ? What is the probability that Fred is always ahead of Bert in the voting ?

2. For a simple random walk, with probabilities p and q, write down the probability that the object reaches $2a$ ($a > 0$) after $2n$ steps, starting from 0. If $p = q = \frac{1}{2}$, what is the probability that it does so without returning to 0 in between ?

3. For a simple random walk with probabilities p and q, write down the probability that the object goes from point $2a$ to point $2b$ ($a, b > 0$) in $2n$ steps.

4. Let $A_{2k, 2n}$ be the probability that in the first $2n$ steps of a random walk with $p = q = \frac{1}{2}$ the walk is at 0 for the last time at $2k$. Show that
$$A_{2n, 2k} = u_{2k} u_{2n-2k} \quad .$$

5. If S_{2n} is the position of the object after $2n$ steps, write down expressions for the mean and variance of S_{2n}, for general p and q. For $p = q = \frac{1}{2}$, use a Normal approximation to compute the probability that the random walk is more than 20 units from the origin after 50 steps.

11.8 Computer Projects

1. For a simple random walk of length $2n$, let x be the proportion of time that the object is above the axis. Adapt the simulation program to compute an approximate probability distribution for x.

2. Let us define a general random walk so that a step of size k is taken with probability p_k each time (for the simple random walk, $p_1 = p$, $p_{-1} = q$). Adapt the simulation program to deal with the case of a general random walk.

CHAPTER 12

Gambler's Ruin

12.1 Absorption Probabilities

Suppose two gamblers, A and B, are betting against each other, and A wins £1 from B each time with probability $\frac{1}{2}$, and B wins £1 from A with probability $\frac{1}{2}$. If A has £a to start with, and B has £b, what is the probability that A will lose all his money to B ? This is equivalent to the following random walk problem:

Let S_n be the amount B has won from A in n turns. Then if $S_n = a$ before $S_n = -b$, A goes broke, whereas if $S_n = -b$ before $S_n = a$, B is the one to go broke. We can consider there to be "absorbing barriers" at a and $-b$, and imagine that the random walk will stop or be absorbed as soon as it reaches one or other of the barriers.

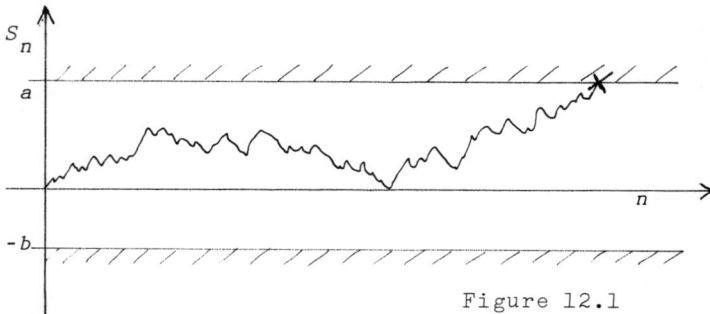

Figure 12.1

182

It is interesting to see how to obtain the probability that the random walk is absorbed at *a* (i.e. that A goes broke before B does). Let *E* be the event that the random walk is absorbed at *a*, and suppose that at time 0 the position S_0 could have any arbitrary value *i*. Define:

$$f(i) = P(E|S_0=i),$$

i.e. the probability of absorption at *a* starting from position *i*. It is fairly obvious that $f(-b) = 0$ and $f(a) = 1$.

To obtain a formula for $f(i)$, we decompose it with respect to the position after the first step (this kind of argument was used in the previous chapter).

$$f(i) = \sum_j P(E \text{ and } S_1 = j | S_0 = i) ,$$

where the summation is over all positions *j* that can be reached after one step. We can say that

$$P(E \text{ and } S_1 = j|S_0 = i) = P(E|S_1 = j \text{ \& } S_0 = i)$$
$$\times P(S_1 = j|S_0 = i) .$$

(If this is not immediately obvious, remember from chapter 4 that $P(A \text{ and } B) = P(A|B) \times P(B)$).

From the structure of the random walk, it is clear that once we know $S_1 = j$, the information that $S_0 = i$ is irrelevant. So $P(E \text{ and } S_1 = j|S_0 = i) = P(E|S_1 = j) \times P(S_1 = j|S_0 = i)$.

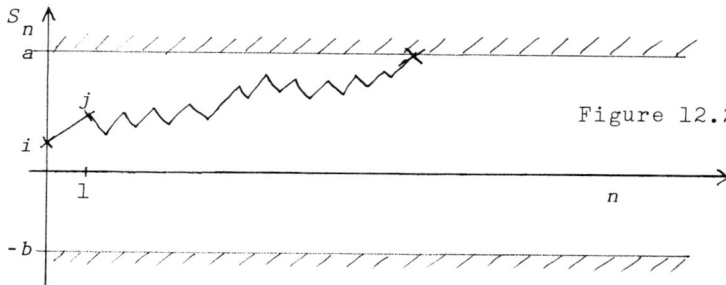

Figure 12.2

When discussing absorption, we do not care about the *time* at which it occurs - any time may be called 0. So $P(E|S_1 = j) = P(E|S_0 = j) = f(j)$.
Therefore,

$$P(E \text{ and } S_1 = j|S_0 = i) = f(j)P(S_1 = j|S_0 = i)$$

and

$$f(i) = \sum_j f(j)P(S_1 = j | S_0 = i) \quad .$$

This relationship is generally true for any random walk with absorbing barriers. If we consider the special case of the simple random walk with $p = q = \frac{1}{2}$, then we can see that the only possible values of j are $i-1$ and $i+1$.

So $f(i) = \frac{1}{2}f(i-1) + \frac{1}{2}f(i+1)$.

An equation of this type is known as a *difference equation*, and to solve it we need to find an expression for $f(i)$ which satisfies the equation for all values of i. Intuitively, the equation seems to be saying that $f(i)$ is the average of $f(i-1)$ and $f(i+1)$, which would lead us to expect a linear form for $f(i)$, something like:

$$f(i) = c_1 i + c_2 ,$$

where c_1 and c_2 are constants.
Substituting this in the difference equation, we get:

R.H.S. $= \frac{1}{2}c_1(i-1) + \frac{1}{2}c_2 + \frac{1}{2}c_1(i+1) + \frac{1}{2}c_2$

$\qquad = c_1 i + c_2 = f(i) = $ L.H.S.

So this is a general solution of the difference equation for any values of c_1 and c_2 - the next problem is to find the correct values of these constants for our particular problem. To do this, we use the so-called *boundary conditions*, that is to say the facts that $f(-b) = 0$ and $f(a) = 1$. These give:

$$-bc_1 + c_2 = 0$$
and $\qquad ac_1 + c_2 = 1$.

Solving these, we get:

$$c_1 = 1/(a+b) \quad \text{and} \quad c_2 = b/(a+b) \quad .$$

Finally we have:

$$f(i) = \frac{i + b}{a + b} \quad .$$

So if we start from $i = 0$, the probability that A goes broke before B is $b/(a+b)$. For example, if A starts with £100, and B is the bank with £100,000 , then the required probability is $100,000/100,100 = 0.9990$.

The moral is: Don't try to break the bank, unless you have more money than they do.

12.2 Expected Time to Absorption

Having computed this probability function, the next problem is to work out how long it will be before either barrier is reached. Let T be the number of steps before one or other barrier is reached, and define

$$\tau(i) \;=\; E(T \,|\, S_0 = i) \;.$$

Again, to compute $\tau(i)$ we must decompose with respect to the position after one step has been taken.

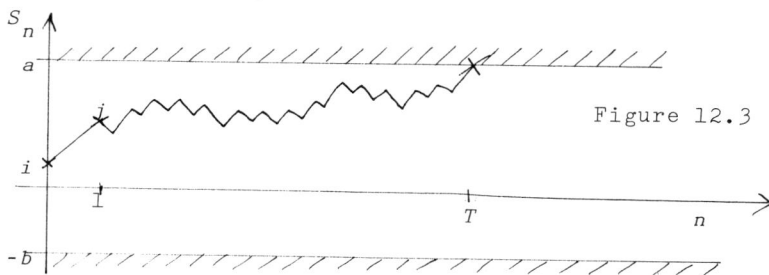

Figure 12.3

$$\tau(i) \;=\; \sum_j E(T \,|\, S_1 = j, S_0 = i) \times P(S_1 = j \,|\, S_0 = i) \;,$$

where $E(T \,|\, S_1 = j, S_0 = i)$ is the expected time to absorption from $S_0 = i$, given that the first step is to position j. It is fairly clear that

$$E(T \,|\, S_1 = j, S_0 = i) \;=\; E(T \,|\, S_1 = j) \;=\; E(T \,|\, S_0 = j) + 1$$
$$=\; \tau(j) + 1 \;.$$

So we get:

$$\tau(i) \;=\; \sum_j (\tau(j) + 1)\, P(S_1 = j \,|\, S_0 = i)$$
$$=\; \sum_j \tau(j)\, P(S_1 = j \,|\, S_0 = i) \;+\; \sum_j P(S_1 = j \,|\, S_0 = i)$$
$$=\; \underline{\sum_j \tau(j)\, P(S_1 = j \,|\, S_0 = i) \;+\; 1} \;.$$

In the special case of the simple random walk with $p = q = \frac{1}{2}$, this reduces to

$$\tau(i) \;=\; \tfrac{1}{2}\tau(i+1) + \tfrac{1}{2}\tau(i-1) + 1 \;.$$

This difference equation does not have a linear general solution, so we shall try a quadratic solution of the form:

$$\tau(i) = -i^2 + c_1 i + c_2 \quad .$$

$$\text{R.H.S.} = -\tfrac{1}{2}(i+1)^2 + \tfrac{1}{2}c_1(i+1) + \tfrac{1}{2}c_2$$

$$- \tfrac{1}{2}(i+1)^2 + \tfrac{1}{2}c_1(i-1) + \tfrac{1}{2}c_2 + 1$$

$$= -i^2 + c_1 i + c_2 = \text{L.H.S.}$$

So this general solution satisfies the difference equation. To find c_1 and c_2, we make use of the boundary conditions $\tau(a) = 0$ and $\tau(-b) = 0$, giving

$$-a^2 + c_1 a + c_2 = 0$$
$$-b^2 - c_1 b + c_2 = 0 \quad .$$

So $c_1(a+b) = a^2 - b^2 \Rightarrow c_1 = a - b$,

and $c_2 = (a-b)b + b^2 = ab$.

Therefore,
$$\tau(i) = -i^2 + (a-b)i + ab$$
$$= (a-i)(i+b) \quad .$$

In particular, for $i = 0$, the expected time before absorption $= ab$. So, in our previous example, with $a = 100$ and $b = 100,000$, the expected number of games before one or other player was broke would be $100 \times 100,000 = 10,000,000$. Playing 10 games per minute, this would take 99 weeks, 1 day, 10 hours and 40 minutes, playing non-stop.

Examples:

1) A pseudo-random number generator is supposed to produce values uniformly distributed in $[0,1]$. Each time it produces a number, it is tested to see if it is greater than or less than $\tfrac{1}{2}$. Once the generator has produced 100 more of one type than the other, a warning message will be printed and the program stopped. If it has already produced 30 values $> \tfrac{1}{2}$ and 20 values $< \tfrac{1}{2}$, what is the expected number of further values which will be generated, if the program is in fact random ? What is the probability that it will stop with a message that it is biased to low values ?

Model this as a simple random walk with S_n = the difference between the number of large values and the number of small values after n further values have been generated.

$S_0 = 30 - 20 = 10 = i$.

$a = b = 100$.

$f(i) = (i+b)/(a+b) = (10+100)/200 = 0.55$.

P(Stops at lower barrier) $= 1 - f(i) = \underline{0.45}$.

$\tau(i) = (a-i)(i+b) = 90 \times 110 = \underline{9900}$.

2) A random walk starts from point i at time 0. At each step, it stays where it is with probability p, or moves up one unit with probability $q = 1-p$. Find a formula for $\tau(i)$, the expected number of steps before reaching a barrier at a.

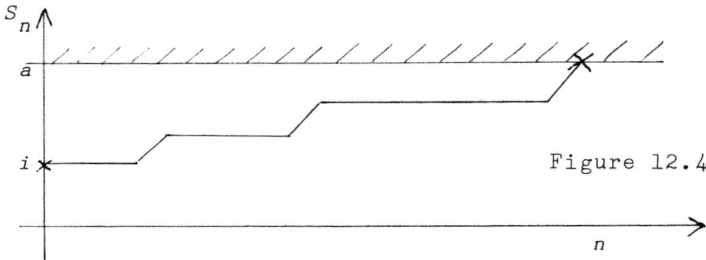

Figure 12.4

Using the relationship:

$$\tau(i) = \sum_j \tau(j) \, P(S_1 = j | S_0 = i) + 1 ,$$

we get

$$\tau(i) = p\tau(i) + q\tau(i+1) + 1$$

$$\Rightarrow \quad (1-p)\tau(i) = q\tau(i+1) + 1$$

$$\Rightarrow \quad \tau(i) = \tau(i+1) + \frac{1}{q} .$$

A general solution of this is clearly

$$\tau(i) = c - \frac{i}{q} , \quad \text{where } c \text{ is a constant.}$$

To find c, set $\tau(a) = 0$, to get $c = a/q$.

So the formula is:

$$\tau(i) = \underline{\frac{a-i}{q}} .$$

12.3 The General Random Walk

We can generalise the simple random walk by allowing the steps each time to be general random variables, instead of just ± 1 with probabilities p and q. In other words, if X_k is the distance moved at the kth step, we write

$$S_n = \sum_{k=1}^{n} X_k \quad \text{as the position after } n \text{ steps.}$$

If each X_k is independent and identically distributed, with mean μ and variance σ^2, then clearly

$$E(S_n) = n\mu \quad \text{and} \quad \text{Var}(S_n) = n\sigma^2 .$$

For the Gambler's Ruin problem in this case, we can use the arguments by decomposition to develop difference equations for the probability of and expected time to absorption.

<u>Discrete case</u>: If $p_\ell = P(X_k = \ell)$, then we have

$$f(i) = \sum_\ell f(i+\ell) \, p_\ell \quad \text{and} \quad \tau(i) = \sum_\ell \tau(i+\ell) \, p_\ell + 1 \quad .$$

<u>Continuous case</u>: If X_k is a continuous random variable with probability density function $q(x)$, and $f(y)$ is the probability of absorption at a starting from position y, then

$$f(y) = \int_{-\infty}^{\infty} f(x+y) q(x) \, dx$$

$$\text{and} \quad \tau(y) = \int_{-\infty}^{\infty} \tau(x+y) q(x) \, dx + 1 \quad .$$

In general, however, these equations are going to be rather difficult to solve, so to deal with the general random walk we shall use an approximation. The Central Limit Theorem (see chapter 4) tells us that, as n increases, the position S_n will tend to be Normally distributed. We can use the Normal distribution to solve certain special cases of the Gambler's Ruin problem, in particular when we have just one barrier.

12.4 The Insurance Problem

An insurance company starts with capital £a. Each month the difference between its income and the amount it must pay out is £X, where X is a random variable with mean μ (> 0) and variance σ^2. What is the probability that the insurance company will ever go broke ?

Diagrammatically, we can represent the situation as follows:

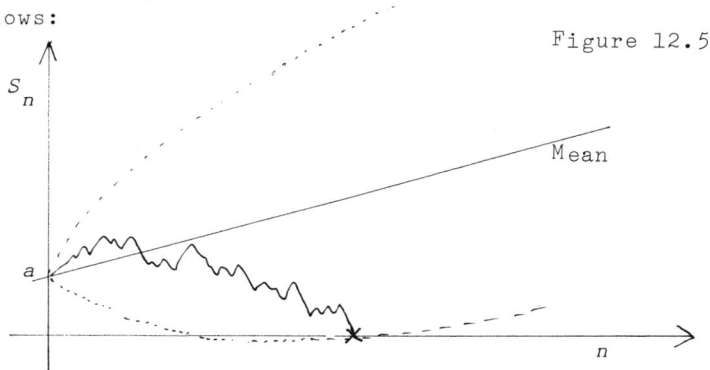

Figure 12.5

The dotted lines represent the $(1-\alpha)$ confidence limits for S_n as a function of n. Using a Normal approximation, we can write down the value for the lower limit as follows:

$$E(S_n) \quad = \quad a + n\mu$$
$$\text{Var}(S_n) \quad = \quad n\sigma^2 \quad \Rightarrow \quad \text{s.d.} \quad = \quad \sqrt{n}\,\sigma \ .$$
$$\text{So the lower limit} \quad = \quad a + n\mu - z_\alpha \sigma \sqrt{n} \ .$$

If this lower limit ever reaches 0, there is a probability of at least $\alpha/2$ of being below it, and thus of going bankrupt. To find n at this point, we set

$$a + n\mu - z_\alpha \sigma \sqrt{n} \quad = \quad 0$$
$$\Rightarrow \quad n^2\mu^2 + n(2a\mu - z_\alpha^2 \sigma^2) + a^2 \quad = \quad 0$$
$$= \quad n \quad = \quad \frac{(z_\alpha^2 \sigma^2 - 2a\mu) \pm \sqrt{(z_\alpha^4 \sigma^4 - 4a\mu z_\alpha^2 \sigma^2)}}{2\mu^2}$$

This formula only has a solution for *n* if
$$z_\alpha^2 \sigma^2 \geq 4a\mu .$$
To find the critical value of z_α for which the lower limit just touches 0, set
$$z_\alpha = \sqrt{4a\mu/\sigma^2} .$$

Example: The initial capital is £5m, and the net income has mean £1m per month and standard deviation £2m.

For $\alpha = 5\%$, $z_\alpha = 1.96$ and we get the following upper and lower limits for S_n:

n	$E(S_n)$	Lower limit	Upper limit
1	6	2.08	9.92
2	7	1.46	12.54
3	8	1.21	14.79
4	9	1.16	16.84
5	10	1.24	18.76
6	11	1.40	20.60
9	14	2.24	35.52

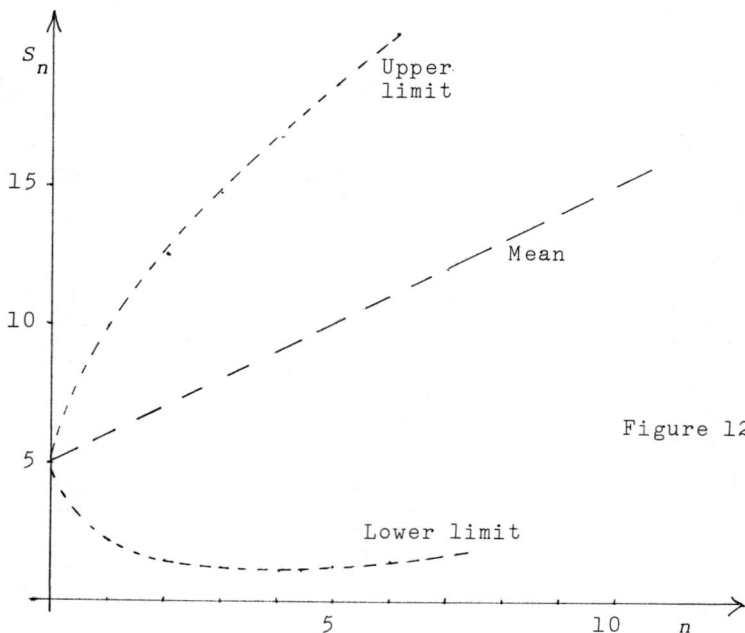

Figure 12.6

Thus there is a less than $2\frac{1}{2}\%$ chance of bankruptcy. To find the limiting value of α, set $z_\alpha = \sqrt{4a\mu/\sigma^2} = 2.24$. So $\alpha = 0.025$, and the probability of bankruptcy is about $1\frac{1}{4}\%$.

12.5 Exercises

1. Consider the Gambler's Ruin problem for a simple random walk with $p \neq q$, and barriers at a and $-b$. Form the difference equation for $f(i)$, the probability of reaching the barrier at a starting from point i. Show that a general solution is:
$$f(i) = \alpha(q/p)^i + \beta .$$
   Determine the values of α and β.

   A small filing system has a maximum capacity of 10 files. Every time the system is accessed, either an existing file is deleted, with probability 1/3, or a new file is created, with probability 2/3. If at present there are 4 files, calculate the probability of the filing system ever being empty before it is full.

2. How long would you expect to have to toss a coin, at one toss per minute, before you got 100 more heads than tails or tails than heads ?

3. A random walk with absorbing barriers at a and $-b$ takes a positive step with probability p $(< \frac{1}{2})$, and a negative step with probability p, and stays constant with probability $1-2p$. Find a formula for the expected time to absorption.

   A computer system has 10 users at present, and a maximum capacity of 30. Each minute there is a probability of 0.1 of a user leaving, and the same probability of new one joining. What is the expected time before the system either is empty or reaches its maximum capacity ?

4. An oil exploration company has a working capital of £10m.
    Each exploration well it drills costs £1m, and has the
    following probabilities for oilfields of various values
    being discovered:

Value of field (£m)	Probability
0	0.8
5	0.15
10	0.05

    Estimate approximately the probability that the company
    will go bankrupt.

5. A random walk has an absorbing barrier at a, and takes a
    step of +1 with probability p and a step of +2 with prob-
    ability q (= 1-p). Derive a formula for the expected time
    to absorption. Generalise your result to any random walk
    which only takes non-negative steps - i.e. step 0 with
    probability p_0, step +1 with probability p_1, etc.

12.6 Computer Projects

1. Develop a program to simulate the Gambler's Ruin problem
    with a simple random walk and general p and q. Use it to
    compare with the analytical results produced in Exercise 1
    above.

2. Expand the program to simulate the general random walk
    (with discrete variables) and use it to produce absorption
    probabilities and expected time to absorption.

CHAPTER 13

Markov Chains

13.1 Definitions

Often in statistics it is useful to be able to assume that a series of random variables are independent - knowledge of the value of one of them gives no information about the others. For example, we normally assume that a sample of n values is independent and identically distributed. However, it is not always possible to make this strong assumption. It would not be right to assume that the rainfall day by day was independent, or that daily stock market prices were independent. This is the reason for considering a weaker property of series of ordered random variables, called the *Markovian property*.

Suppose we have a sequence of random variables $Z_1, Z_2, \ldots Z_n$. If they are independent, one way of writing this is to say that $\quad P(Z_n = z | Z_1, Z_2, \ldots Z_{n-1}) = P(Z_n = z)$.
In other words, the extra information about the values of Z_1 up to Z_{n-1} tells us nothing about the probability for Z_n. Clearly we can extend this idea a bit and write
$$P(Z_n = z | Z_1, Z_2, \ldots Z_{n-1}) = P(Z_n = z | Z_{n-1}).$$
In other words, the probability distribution for Z_n depends on the value of Z_{n-1}, but not on any preceding values. This is the "next step" from total independence - dependence only on the preceding value in the series. A sequence of random

193

variables with the above property for all values of n is
called a *Markov process*. If we make the further assumption
that the probability distributions for the random variables
are all the same, i.e. $P(Z_k = z)$ is the same for all k, then
we have a *Markov chain*. We may therefore define the transition
probability: $p(z,x) = P(Z_k = z | Z_{k-1} = x)$, which is the
same for all k.

The underlying random variable Z_k for the Markov chain
may be either discrete or continuous. From here on, we shall
consider only the case when it is discrete, and its possible
values may be labelled by a discrete set of *states* whose index
is $i = 1,2,3, \ldots m$ or ∞. We shall mainly discuss Markov
chains by reference to their *one-step transition probability
matrix* P, whose elements are:

$$p_{ij} = P(Z_k = j | Z_{k-1} = i)$$
$$= P(\text{Going from } i \text{ to } j \text{ in one step}).$$

We may also define the *n-step transition probability matrix*
$P^{(n)}$, whose elements are:

$$p_{ij}^{(n)} = P(Z_k = j | Z_{k-n} = i)$$
$$= P(\text{Going from } i \text{ to } j \text{ in } n \text{ steps}).$$

Mathematically, we can describe the properties of a
Markov chain using these transition probability matrices.
However, if we want a graphical impression of a Markov chain,
we may produce a diagram in which each state is represented
by a numbered circle, and an arrow joins one state to another
if the probability of getting from one to the other in one
step is greater than zero.

Example: A frog lives in a lily pond with 5 lily leaves, and
each day he either moves to a new leaf or stays on
his present one. Treating the 5 leaves as the 5
states of a Markov chain, the one-step matrix is:

		To				
		1	2	3	4	5
P =	1	0	1/2	1/2	0	0
	2	1/3	0	2/3	0	0
From	3	3/4	0	0	1/4	0
	4	0	0	0	0	1
	5	0	0	0	1	0

This may be represented by the following diagram:

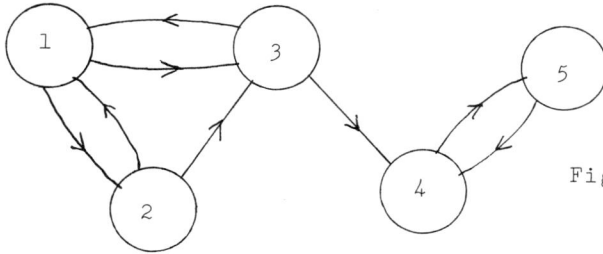

Figure 13.1

13.2 Classification of Markov Chain States

There are various ways in which the states of a Markov chain may be described.

13.2.1 COMMUNICATING/DISJOINT

For any pair of states i and j, we can write $i \to j$ if there is some number of steps n such that $p_{ij}^{(n)} > 0$. In other words, if it is possible to get from i to j in n steps. Similarly, if there is a number of steps r such that $p_{ji}^{(r)} > 0$, then we write $j \to i$. If both $i \to j$ and $j \to i$ then we say that i and j are communicating states, and write $i \leftrightarrow j$. In the previous example, states 2 and 3 are communicating (you can get from 2 to 3 in one step, and from 3 to 2 in 2 steps). However, 3 and 4 are not communicating, because although $3 \to 4$, there is no way of getting from 4 to 3.

Two states i and j are disjoint if you cannot get from i to j in any number of steps, or from j to i. For example:

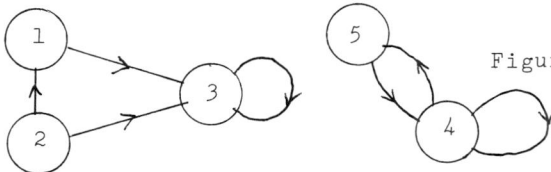

Figure 13.2

In this case states 1 and 5 are disjoint.

13.2.2 PERIODIC

If some of the states of a Markov chain can be divided into d separate groups, so that if i and j are in the same group $p_{ij}^{(n)} > 0$ only if n is divisible by d, then these states are periodic with period d. For example:

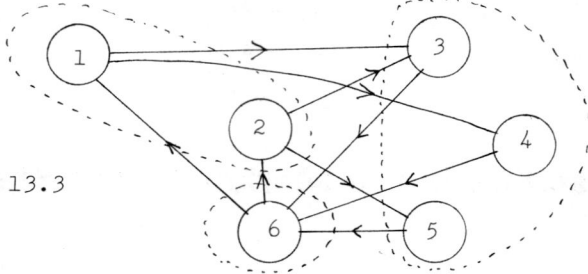

Figure 13.3

The first group is states 1 and 2, the second is 3, 4 and 5, and the third is just state 6. The period is 3 steps, because whatever group of states you are in, you will get back to the same group (not necessarily the same state) every 3 steps.

In the frog example, states 4 and 5 are periodic with period 2 - the frog continuously goes from 4 to 5 and back again once it reaches these states.

13.2.3 ABSORBING

A state i is absorbing if it is impossible to leave it once you enter it; $p_{ii} = 1$ and $p_{ij} = 0$ if $i \neq j$. This obviously implies that, for all n: $p_{ii}^{(n)} = 1$ and $p_{ij}^{(n)} = 0$ if $i \neq j$. A Markov chain is absorbing if from every state you can reach an absorbing state.

Example: In our frog example, the fifth lily leaf is in fact a disguised trap beneath which lurks the Lesser Spotted Markov Frog Eater. When the frog reaches here, it becomes supper. This may be modelled by means of a small change to the one-step transition probability matrix.

	1	2	3	4	5
1	0	1/2	1/2	0	0
2	1/3	0	2/3	0	0
3	3/4	0	0	1/4	0
4	0	0	0	0	1
5	0	0	0	0	1

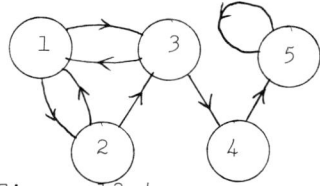

Figure 13.4

State 5 is now absorbing, and the chain is an absorbing chain. The frog is ultimately doomed to be swallowed.

13.2.4 RECURRENT/TRANSIENT

Define for any state i:
$$r_i = P(\text{Eventually returns to } i \,|\, \text{Starts in } i).$$

If $r_i = 1$, the state is called recurrent, because we continually return to it. If $r_i < 1$, the state is called transient, because we will eventually leave it and never return. Thus the state is only visited a finite number of times.

In our first example, the frog in the lily pond (see Figure 13.1), states 1, 2 and 3 are all transient. That is to say, if the frog starts in one of these states, it will eventually jump from 3 to 4, and from state 4 there is no possibility of a return to states 1, 2 or 3. States 4 and 5 are recurrent. In the revised frog example (Figure 13.4), not only are states 1, 2 and 3 transient, but so is state 4. From there the frog can only go to state 5, which is absorbing, and there is no return to 4 at all.

Let us consider another Markov chain, and try to identify the transient and recurrent states.

$$P = \begin{array}{c} \\ 1 \\ 2 \\ 3 \\ 4 \\ 5 \\ 6 \end{array} \begin{array}{cccccc} 1 & 2 & 3 & 4 & 5 & 6 \\ \left(\begin{array}{cccccc} 1/2 & 0 & 0 & 0 & 1/4 & 1/4 \\ 0 & 1/3 & 0 & 2/3 & 0 & 0 \\ 0 & 1/5 & 0 & 4/5 & 0 & 0 \\ 0 & 1/6 & 1/3 & 1/2 & 0 & 0 \\ 0 & 1/8 & 0 & 0 & 0 & 7/8 \\ 3/8 & 0 & 3/8 & 0 & 0 & 1/4 \end{array} \right) \end{array}$$

Diagrammatically:

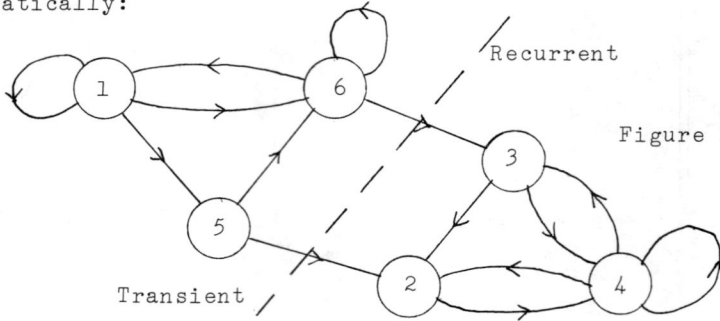

Figure 13.5

It is not instantly clear from the one-step transition probability matrix (P) which states are transient and which recurrent. Inspection of the diagram shows that between the group of states 1, 5 and 6, and the group consisting of states 2, 3 and 4, there is only one direction of transition. Thus the former are transient, since once we get into the latter group we cannot move back into the former group. Another way of expressing this is to note that a line may be drawn across the diagram dividing the two groups of states, and that across this line the arrows only go in one direction.

13.3 Transition Probability Matrices

We have seen that we can define the behaviour of a Markov chain in terms of its n-step transition probability matrices $P^{(n)}$, whose elements are:

$$p_{ij}^{(n)} = \text{P(Going from state } i \text{ to state } j \text{ in } n \text{ steps)} .$$

Notice that the probability of going from state i to any other state in n steps $= p_{i1}^{(n)} + p_{i2}^{(n)} + \ldots + p_{im}^{(n)} = 1.0$.

Therefore, any transition probability matrix must obey the following rules:

1. It must be square (number of rows = number of columns).
2. All its elements must be ≥ 0 and ≤ 1.
3. Each row must sum to 1.0 .

Any matrix with these properties is called a *stochastic*

matrix. Each Markov chain has an infinite series of such
n-step transition probability matrices associated with it, as
well as the one-step matrix P. It would be useful to be able
to derive the n-step matrices from the one-step matrix P.

The elements of the matrix $P^{(n)}$ are:

$$p_{ij}^{(n)} = P(\text{Going from } i \text{ to } j \text{ in } n \text{ steps}).$$

Let us suppose that at the $(n-1)$th step we were in state k.
The probability of this is equal to:

$$P(\text{Going from } i \text{ to } k \text{ in } n-1 \text{ steps}) \times P(k \text{ to } j \text{ in } 1 \text{ step})$$

$$= p_{ik}^{(n-1)} \cdot p_{kj}^{(1)} \ .$$

To get the total probability for n steps we need to sum over
all possible intermediate states k to get:

$$P(i \text{ to } j \text{ in } n \text{ steps}) = \sum_{k=1}^{m} p_{ik}^{(n-1)} \cdot p_{kj}^{(1)} = p_{ij}^{(n)} \ .$$

So, for $n = 2$ we get:

$$p_{ij}^{(2)} = \sum_{k=1}^{m} p_{ik}^{(1)} \cdot p_{kj}^{(1)} \ .$$

From chapter 3 we can see that this just corresponds to the
definition of matrix multiplication. Therefore,

$$P^{(2)} = P^{(1)} \cdot P^{(1)} = P \cdot P = P^2 \ .$$

Similarly,

$$P^{(3)} = P^{(2)} \cdot P^{(1)} = P^2 \cdot P = P^3 \ ,$$

$$P^{(n)} = P^{(n-1)} \cdot P^{(1)} = P^{n-1} \cdot P = P^n \ .$$

So given the one-step matrix P, we can obtain any of the n-step
matrices just by the operation of matrix multiplication.

Example: Frog example -

$$P = \begin{pmatrix} 0 & 1/2 & 1/2 & 0 & 0 \\ 1/3 & 0 & 2/3 & 0 & 0 \\ 3/4 & 0 & 0 & 1/4 & 0 \\ 0 & 0 & 0 & 0 & 1 \\ 0 & 0 & 0 & 1 & 1 \end{pmatrix}$$

$$P^2 = \begin{pmatrix} 0 & 1/2 & 1/2 & 0 & 0 \\ 1/3 & 0 & 2/3 & 0 & 0 \\ 3/4 & 0 & 0 & 1/4 & 0 \\ 0 & 0 & 0 & 0 & 1 \\ 0 & 0 & 0 & 1 & 0 \end{pmatrix} \begin{pmatrix} 0 & 1/2 & 1/2 & 0 & 0 \\ 1/3 & 0 & 2/3 & 0 & 0 \\ 3/4 & 0 & 0 & 1/4 & 0 \\ 0 & 0 & 0 & 0 & 1 \\ 0 & 0 & 0 & 1 & 0 \end{pmatrix}$$

$$= \begin{pmatrix} 13/24 & 0 & 1/3 & 1/8 & 0 \\ 1/2 & 1/6 & 1/6 & 1/6 & 0 \\ 0 & 3/8 & 3/8 & 1/4 & 0 \\ 0 & 0 & 0 & 1 & 0 \\ 0 & 0 & 0 & 0 & 1 \end{pmatrix}$$

$$P^3 = \begin{pmatrix} 1/4 & 13/48 & 13/48 & 1/12 & 1/8 \\ 13/72 & 1/4 & 13/36 & 1/24 & 1/6 \\ 13/32 & 0 & 1/4 & 3/32 & 1/4 \\ 0 & 0 & 0 & 0 & 1 \\ 0 & 0 & 0 & 1 & 0 \end{pmatrix}$$

The following program computes and prints out an n-step transition probability matrix, given a one-step matrix.

```
program nstep(input,output);
{ Program to compute n-step matrix for Markov chain. }
const maxstates = 10;
type stochastic = array[1..maxstates] of real;
     posint = 0..maxint;
var  p,pn : stochastic;
     i,j,step,nstates,n : posint;
     matOK : boolean;

procedure checkit(var p : stochastic; m : posint;
                  var OK : boolean);
{ Procedure to check that p is a valid stochastic matrix. }
var i,j : posint;
    sum : real;
begin
  OK := true;
  for i := 1 to m do
  begin
    sum := 0;
    for j := 1 to m do
    begin
      sum := sum + p[i,j];
      if (p[i,j] < 0) or (p[i,j] > 1.0) then OK := false;
    end;
    if  abs(sum-1.0) > 1.0e-6 then OK := false;
  end;
end;

procedure matmult(var pmult,p : stochastic; m : posint);
{ Matrix multiplication routine. }
var i,j,k : posint;
    pdum : stochastic;
begin
  for i := 1 to m do
    for j := 1 to m do pdum[i,j] := pmult[i,j];
  for i := 1 to m do
    for j := 1 to m do
    begin
```

200

```
            pmult[ i,j ] := 0.0;
            for k := 1 to m do
               pmult[ i,j ] := pmult[ i,j ] + pdum[ i,k ]*p[ k,j ];
         end;
   end;

   procedure matprint(var p : stochastic; m : posint);
   { Matrix printing routine. }
   var i,j : posint;
   begin
      writeln;
      for i := 1 to m do
      begin
         for j := 1 to m do write(p[ i,j ]:7:4);
         writeln;
      end;
   end;

   { Main routine. }
   begin
      read(nstates,n);
      for i := 1 to nstates do
         for j := 1 to nstates do read(p[ i,j ]);
      writeln; writeln('One-step probability matrix -');
      matprint(p,nstates);
      checkit(p,nstates,matOK);
      if matOK then
      begin
         for i := 1 to nstates do
            for j := 1 to nstates do pn[ i,j ] := p[ i,j ];
         for step := 2 to n do matmult(pn,p,nstates);
         writeln;
         writeln(n:3,'-step probability matrix -');
         matprint(pn,nstates);
      end
      else writeln('Matrix is not stochastic');
   end.
```

Output from an example run:

One-step probability matrix -

```
0.0000 0.5000 0.5000 0.0000 0.0000
0.3333 0.0000 0.6667 0.0000 0.0000
0.5000 0.2500 0.0000 0.2500 0.0000
0.0000 0.0000 0.0000 0.0000 1.0000
0.0000 0.0000 0.0000 1.0000 0.0000
```

10-step probability matrix -

```
0.1242 0.1112 0.1511 0.3500 0.2635
0.1221 0.1082 0.1483 0.4206 0.2009
0.1032 0.0915 0.1242 0.1820 0.4991
0.0000 0.0000 0.0000 1.0000 0.0000
0.0000 0.0000 0.0000 0.0000 1.0000
```

13.4 Regular Chains and Stable Probabilities

A Markov chain is said to be *regular* is there is some number of steps, n say, such that $p_{ij}^{(n)} > 0$ for all i and j. In other words, if we can get from any state to any other state in n steps. All the states of such a chain must be recurrent. We can show that, for a regular chain, the series of transition probability matrices P, P^2, P^3, P^4 ... tends to a limiting matrix W.

Example: A very simple model of the weather assumes it is in one of three possible states: rain, fine or snow. We also assume that the next day's weather depends only on today's weather, so that we can use a Markov chain model.   Suppose the one-step matrix were:

To

		Rain	Fine	Snow
	Rain	1/2	1/3	1/6
From	Fine	1/3	1/2	1/6
	Snow	2/3	1/6	1/6

By matrix multiplication, the two-step matrix is:

$$P^2 = \begin{pmatrix} 17/36 & 13/36 & 1/6 \\ 4/9 & 7/18 & 1/6 \\ 1/2 & 1/3 & 1/6 \end{pmatrix} = \begin{pmatrix} 0.4722 & 0.3611 & 0.1667 \\ 0.4444 & 0.3889 & 0.1667 \\ 0.5000 & 0.3333 & 0.1667 \end{pmatrix}$$

The three-step matrix is:

$$P^3 = \begin{pmatrix} 101/216 & 79/216 & 1/6 \\ 25/54 & 10/27 & 1/6 \\ 17/36 & 13/36 & 1/6 \end{pmatrix} = \begin{pmatrix} 0.4676 & 0.3657 & 0.1667 \\ 0.4630 & 0.3703 & 0.1667 \\ 0.4722 & 0.3611 & 0.1667 \end{pmatrix}$$

It is fairly clear that the first column is tending to a value just under $\frac{1}{2}$, the second to a value just over 1/3, and the third column is remaining at 1/6. Eventually, the transition matrices will tend to a limit W, with each row the same:

$$W = \begin{pmatrix} \omega_1 & \omega_2 & \omega_3 \\ \omega_1 & \omega_2 & \omega_3 \\ \omega_1 & \omega_2 & \omega_3 \end{pmatrix}$$

These values, ω_1, ω_2 and ω_3, are the *stable probabilities* for the Markov chain, i.e. the long-term probabilities of the chain being in each state, for any arbitrary starting state. These probabilities must sum to 1, and must be such that they are unaffected by taking another step. In other words, the matrix W must be such that $WP = W$. Applying this to our example, we get:

$$\begin{pmatrix} \omega_1 & \omega_2 & \omega_3 \\ \omega_1 & \omega_2 & \omega_3 \\ \omega_1 & \omega_2 & \omega_3 \end{pmatrix} \begin{pmatrix} 1/2 & 1/3 & 1/6 \\ 1/3 & 1/2 & 1/6 \\ 2/3 & 1/6 & 1/6 \end{pmatrix} = \begin{pmatrix} \omega_1 & \omega_2 & \omega_3 \\ \omega_1 & \omega_2 & \omega_3 \\ \omega_1 & \omega_2 & \omega_3 \end{pmatrix}$$

(If you are familiar with the idea of eigenvalues, you will notice that this implies that $(\omega_1, \omega_2, \omega_3)$ is an eigenvector of P with eigenvalue 1).

To find the exact values of ω_1, ω_2, and ω_3, we need to convert the above matrix equation into a set of simultaneous equations. In fact, multiplying out the left-hand side we get 9 equations, but only 3 of these are different. These are:

$$\frac{1}{2}\omega_1 + \frac{1}{3}\omega_2 + \frac{2}{3}\omega_3 = \omega_1 \quad \text{(Column 1)}$$

$$\frac{1}{3}\omega_1 + \frac{1}{2}\omega_2 + \frac{1}{6}\omega_3 = \omega_2 \quad \text{(Column 2)}$$

$$\frac{1}{6}\omega_1 + \frac{1}{6}\omega_2 + \frac{1}{6}\omega_3 = \omega_3 \quad \text{(Column 3)} \ .$$

Re-arranging, we get:

$$-\frac{1}{2}\omega_1 + \frac{1}{3}\omega_2 + \frac{2}{3}\omega_3 = 0$$

$$\frac{1}{3}\omega_1 - \frac{1}{2}\omega_2 + \frac{1}{6}\omega_3 = 0$$

$$\frac{1}{6}\omega_1 + \frac{1}{6}\omega_2 - \frac{5}{6}\omega_3 = 0 \ .$$

We have three equations in three unknowns, but unfortunately they do not give us a unique solution because of the zeroes on the right-hand side. The way to get round this problem is to replace one of the three equations by the condition that the stable probabilities must sum to 1. So we end up with:

$$-\tfrac{1}{2}\omega_1 + \tfrac{1}{3}\omega_2 + \tfrac{2}{3}\omega_3 = 0 \qquad (1)$$

$$\tfrac{1}{3}\omega_1 - \tfrac{1}{2}\omega_2 + \tfrac{1}{6}\omega_3 = 0 \qquad (2)$$

$$\omega_1 + \omega_2 + \omega_3 = 1 \qquad (3) \quad .$$

$$4\times(2) - (1) \;\Rightarrow\; \tfrac{11}{6}\omega_1 - \tfrac{7}{3}\omega_2 = 0 \qquad (4)$$

$$6\times(2) - (3) \;\Rightarrow\; \omega_1 - 4\omega_2 = -1 \qquad (5)$$

$$6\times(4) - \tfrac{7}{2}\times(5) \;\Rightarrow\; \tfrac{15}{2}\omega_1 = \tfrac{7}{2} \quad .$$

Therefore,
$$\omega_1 = 7/15 = 0.4667$$
$$\omega_2 = 11/30 = 0.3667$$
$$\omega_3 = 1/6 = 0.1667 \quad .$$

In other words, on average it will rain 46.67% of the time, be fine 36.67% of the time, and snow 16.67% of the time.

Another way of solving this problem, which will be more useful for larger problems with more states and especially for computer implementation, involves matrix inversion. Let $\underset{\sim}{\omega}$ be the vector of stable probabilities, and our basic equation becomes

$$\underset{\sim}{\omega}'P = \underset{\sim}{\omega}'$$

or
$$\underset{\sim}{\omega}'(P-I) = \underset{\sim}{0}' \, ,$$

where I is a unit matrix and $\underset{\sim}{0}$ is a vector of zeroes. We also have $\underset{\sim}{\omega}'\underset{\sim}{1} = 1$, where $\underset{\sim}{1}$ is a vector of 1's. This is the vector form of the equation $\omega_1 + \omega_2 + \omega_3 = 1$.

Replace the last column of $P-I$ by a column of 1's, and call the resulting matrix R. We now have:

$$\underset{\sim}{\omega}'R = \underset{\sim}{b}' , \quad \text{where} \quad \underset{\sim}{b}' = (0,0,1) \quad .$$

The solution is $\underset{\sim}{\omega}' = \underset{\sim}{b}'R^{-1}$,

i.e. $\underset{\sim}{\omega}'$ is the last row of the R^{-1} matrix.

In our example, $R = \begin{Bmatrix} -1/2 & 1/3 & 1 \\ 1/3 & -1/2 & 1 \\ 2/3 & 1/6 & 1 \end{Bmatrix}$

Computing the inverse matrix,

$$R^{-1} = \begin{pmatrix} -4/5 & -1/5 & 1 \\ 2/5 & -7/5 & 1 \\ 7/15 & 11/30 & 1/6 \end{pmatrix} = \begin{pmatrix} -0.8 & -0.2 & 1.0 \\ 0.4 & -1.4 & 1.0 \\ 0.4667 & 0.3667 & 0.1667 \end{pmatrix}$$

The last row gives the same result as before. This matrix
technique will work for any size Markov chain - the dimensions
of the matrices and vectors concerned will correspond to the
number of states in the chain.

13.5 Analysis of Absorbing Chains

An absorbing chain is one in which it is possible to get
to at least one absorbing state from every non-absorbing state.
For example:

$$P = \begin{pmatrix} 0 & 1/2 & 1/2 & 0 & 0 \\ 1/3 & 1/3 & 0 & 1/3 & 0 \\ 1/4 & 1/2 & 0 & 0 & 1/4 \\ 0 & 0 & 0 & 1 & 0 \\ 0 & 0 & 0 & 0 & 1 \end{pmatrix}$$

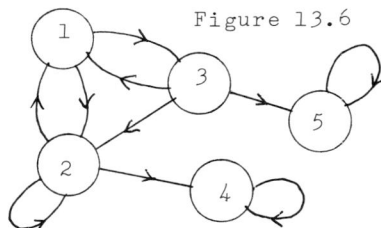

Figure 13.6

When studying absorbing chains, there are two questions
to be answered:

1. How many times do you expect to visit each non-absorbing
   state before being absorbed ?
2. For each non-absorbing state, what are the probabilities
   of ultimate absorption in each absorbing state ?

To answer these questions, we need first to divide our
matrix P into four parts:

	Non-absorbing states	Absorbing states	
Non-absorbing states	Q	R	0 is a zero matrix.
Absorbing states	0	I	I is a unit matrix.

205

In the above example, $Q = \begin{pmatrix} 0 & 1/2 & 1/2 \\ 1/3 & 1/3 & 0 \\ 1/4 & 1/2 & 0 \end{pmatrix}$

The argument to develop the answer to question 1 above is quite instructive, and goes as follows:

Consider two transient states i and j -

P(Going from i to j in n steps) $= q_{ij}^{(n)}$

$= (i,j)$th element of the matrix Q^n.

Now define an "indicator variable" X_n such that

$X_n = 1$ if we go from i to j in n steps

$= 0$ otherwise .

$E(X_n) = q_{ij}^{(n)}$.

Define T_n to be the total number of times we are in state j, having started from state i, during the first n steps.

$$T_n = X_0 + X_1 + \ldots + X_n$$

$$\text{and} \quad E(T_n) = E(X_0) + E(X_1) + \ldots + E(X_n)$$

$$= q_{ij}^{(0)} + q_{ij}^{(1)} + \ldots + q_{ij}^{(n)}$$

$$= (i,j)\text{th element of the matrix sum}$$

$$Q^0 + Q^1 + \ldots + Q^n \quad .$$

If we let n tend to ∞, then T_∞ is the total number of times we ever return to state j, having started in state i. $E(T_\infty)$ is the (i,j)th element of the infinite matrix sum

$$Q^0 + Q^1 + Q^2 + \ldots$$

Define the *fundamental matrix* N to be equal to this matrix sum. In fact it can be shown that $N = (I-Q)^{-1}$.

(Remember that for ordinary numbers, rather than matrices, it is true that $(1-x)^{-1} = 1 + x + x^2 + x^3 + \ldots$).

So $E(T_\infty) = n_{ij}$, which is the (i,j)th element of $(I-Q)^{-1}$.

Example: $\quad Q = \begin{pmatrix} 0 & 1/2 & 1/2 \\ 1/3 & 1/3 & 0 \\ 1/4 & 1/2 & 0 \end{pmatrix}$

$$I-Q = \begin{pmatrix} 1 & -1/2 & -1/2 \\ -1/3 & 2/3 & 0 \\ -1/4 & -1/2 & 1 \end{pmatrix}$$

206

$$(I-Q)^{-1} = \begin{pmatrix} 2 & 9/4 & 1 \\ 1 & 21/8 & 1/2 \\ 1 & 15/8 & 3/2 \end{pmatrix}$$

So, if we start in state 1, we expect to visit state 2 an average of 9/4 times before being absorbed, and so forth. The fundamental matrix N therefore gives the answers to our first question. Notice that all the values of N must be ≥ 0 for a sensible result. Negative values would indicate an error in our calculations.

Our second question concerned the probabilities of absorption in each absorbing state, starting in any non-absorbing state.

Let b_{ik} = P(Absorbed at k | Started at i) be the (i,k)th element of a matrix B. Decompose this probability with respect to an intermediate non-absorbing state j:

P(Going from i to j in 1 step and then absorbed at k)

$$= q_{ij} \cdot b_{jk} \quad .$$

Summing over all non-absorbing states j, we get

$$b_{ik} = \sum_{j} q_{ij} \cdot b_{jk} + r_{ik} \quad .$$

The first term covers all the cases of going to another non-absorbing state before being absorbed. The second term covers the case of going from i to k in one step. In matrix form:

$$B = QB + R$$
$$\text{or} \quad (I-Q)B = R$$
$$\text{or} \quad B = (I-Q)^{-1}R = NR \quad .$$

Example: $N = \begin{pmatrix} 2 & 9/4 & 1 \\ 1 & 21/8 & 1/2 \\ 1 & 15/8 & 3/2 \end{pmatrix} \qquad R = \begin{pmatrix} 0 & 0 \\ 1/3 & 0 \\ 0 & 1/4 \end{pmatrix}$

$$\text{So} \quad B = \begin{matrix} \begin{pmatrix} 3/4 & 1/4 \\ 7/8 & 1/8 \\ 5/8 & 3/8 \end{pmatrix} & \begin{matrix} 1 \\ 2 \\ 3 \end{matrix} \\ \begin{matrix} 4 \quad\; 5 \end{matrix} \end{matrix}$$

If we start in state 1, there is a probability of $\frac{3}{4}$ that we will be absorbed in state 4, and $\frac{1}{4}$ that we will be absorbed in state 5. Notice that each row of B must sum to 1.

207

13.6 Applications of Markov Chain Models

13.6.1 SIMPLE RANDOM WALK

The simple random walk is an example of a Markov chain, with the states being the possible distances from the origin. There are an infinite number of states, and the one-step matrix looks like this:

$$
\begin{array}{c}
 \\
\cdot \\
-2 \\
-1 \\
0 \\
1 \\
2 \\
\cdot \\
\cdot
\end{array}
\begin{array}{c}
\begin{array}{cccccc}
\cdot\cdot\cdot & -2 & -1 & 0 & 1 & 2 & \cdot\cdot\cdot
\end{array} \\
\left(
\begin{array}{cccccc}
\cdot & \cdot & \cdot & \cdot & \cdot \\
\cdots\ \ 0 & \frac{1}{2} & 0 & 0 & 0 & \cdots \\
\cdots\ \ \frac{1}{2} & 0 & \frac{1}{2} & 0 & 0 & \cdots \\
\cdots\ \ 0 & \frac{1}{2} & 0 & \frac{1}{2} & 0 & \cdots \\
\cdots\ \ 0 & 0 & \frac{1}{2} & 0 & \frac{1}{2} & \cdots \\
\cdots\ \ 0 & 0 & 0 & \frac{1}{2} & 0 & \cdots \\
\cdot & \cdot & \cdot & \cdot & \cdot
\end{array}
\right)
\end{array}
$$

This chain is periodic, with period 2 - you always go from odd to even state, and from even to odd. Each state is also recurrent. Random walks with absorbing barriers can be set up as absorbing chains, for example:

$$
\begin{array}{c}
 \\
0 \\
1 \\
2 \\
3 \\
4
\end{array}
\begin{array}{c}
\begin{array}{ccccc}
0 & 1 & 2 & 3 & 4
\end{array} \\
\left(
\begin{array}{ccccc}
1 & 0 & 0 & 0 & 0 \\
\frac{1}{2} & 0 & \frac{1}{2} & 0 & 0 \\
0 & \frac{1}{2} & 0 & \frac{1}{2} & 0 \\
0 & 0 & \frac{1}{2} & 0 & \frac{1}{2} \\
0 & 0 & 0 & 0 & 1
\end{array}
\right)
\end{array}
\qquad
\begin{array}{l}
\text{Absorbing barriers at} \\
\text{0 and 4.}
\end{array}
$$

Analysing this as an absorbing Markov chain, we get:

$$
N = \begin{pmatrix} 3/2 & 1 & 1/2 \\ 1 & 2 & 1 \\ 1/2 & 1 & 3/2 \end{pmatrix}
\qquad
B = \begin{pmatrix} 3/4 & 1/4 \\ 1/2 & 1/2 \\ 1/4 & 3/4 \end{pmatrix}
$$

13.6.2 COMPUTER PROGRAMS

One interesting example of Markov chain modelling is in
the field of computer programming. We may wish to know the
average number of times various sections of code will be
executed, and the probabilities of terminating at various
points in the program. If we assume that at each decision
branch the probabilities of going each way are known and are
independent of past branches, then we may set up a Markov
chain model. For example, consider the following flowchart,
with decision probabilities marked:

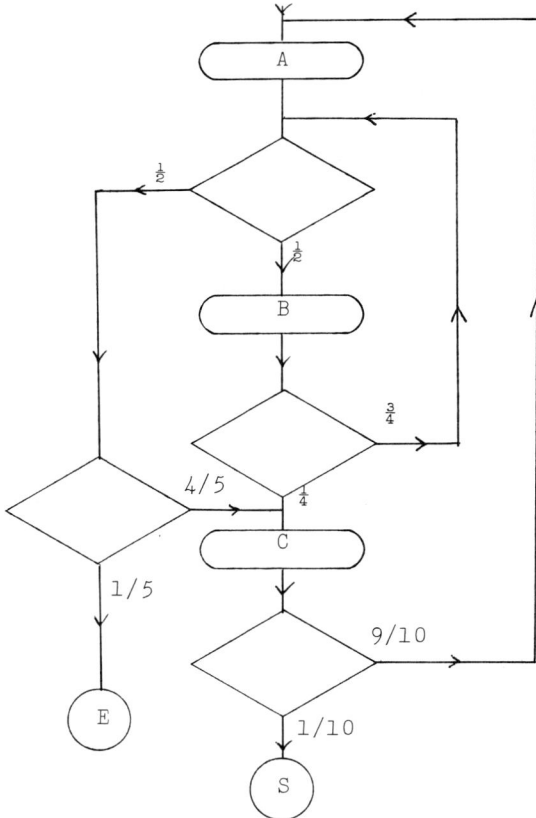

Figure 13.7

E is an error
termination, and S
is a normal
termination.

The Markov chain model of this program has 5 states, and the
following one-step transition probability matrix:

$$
\begin{array}{c}
\quad\;\; A \quad\;\; B \quad\;\; C \quad\;\; E \quad\; S \\
\begin{array}{c} A \\ B \\ C \\ E \\ S \end{array}
\left(
\begin{array}{ccccc}
0 & 1/2 & 2/5 & 1/10 & 0 \\
0 & 3/8 & 11/20 & 3/40 & 0 \\
9/10 & 0 & 0 & 0 & 1/10 \\
0 & 0 & 0 & 1 & 0 \\
0 & 0 & 0 & 0 & 1
\end{array}
\right)
\end{array}
$$

(You should satisfy yourself that this is right, especially the second row).

For the three transient states, A, B and C:

$$
Q = \left(
\begin{array}{ccc}
0 & 1/2 & 2/5 \\
0 & 3/8 & 11/20 \\
9/10 & 0 & 0
\end{array}
\right)
\quad
I\text{-}Q = \left(
\begin{array}{ccc}
1 & -1/2 & -2/5 \\
0 & 5/8 & -11/20 \\
-9/10 & 0 & 1
\end{array}
\right)
$$

$$
N = I\text{-}Q = \left(
\begin{array}{ccc}
4.098 & 3.278 & 3.442 \\
3.246 & 4.196 & 3.606 \\
3.688 & 2.951 & 4.098
\end{array}
\right)
$$

In this case the first row is the most interesting, since we assume that we enter the program at the top. This first row gives the expected number of times each section is executed before the program finishes.

$$
R = \left(
\begin{array}{cc}
1/10 & 0 \\
3/40 & 0 \\
0 & 1/10
\end{array}
\right)
\quad
B = NR = \left(
\begin{array}{cc}
0.656 & 0.344 \\
0.639 & 0.361 \\
0.590 & 0.410
\end{array}
\right)
$$

The elements of B are the probabilities of error or normal termination from each section of code. Of course, this is only a very simple example of this application - more complex programs would require more states to model them.

13.6.3 <u>PHYSICAL PROCESSES - A SIMPLE GAS</u>

Two containers are joined by a fine tube, and the whole system contains three molecules, as in the diagram:

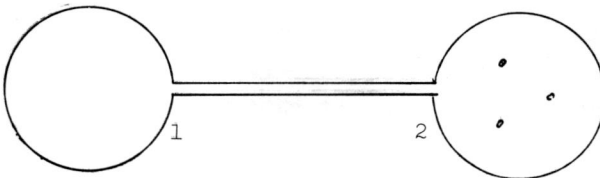

Figure
13.8

Assume that each second each molecule has an independent probability of 0.1 of transferring from its present chamber to the other, and 0.9 of staying where it is. We can model this as a Markov chain, labelling the states according to the number of molecules in the first chamber.

Starting in state 0:

P(From 0 to 0) = P(All molecules stay put) = 0.9^3 = 0.729 .
P(From 0 to 1) = P(2 stay put and 1 comes across)
                = $3 \times 0.9^2 \times 0.1$ = 0.243 .

In a similar way, we can show that p_{02} = 0.027 and p_{03} = 0.001. In fact, the first row probabilities, starting in state 0, form a Binomial probability distribution, and must therefore sum to 1.

Starting in state 1:

p_{10} = P(1 goes and others stay put) = 0.1×0.9^2 = 0.081 .
p_{11} = P(All stay put) + P(1 goes and 1 comes back)
         = $0.9^3 + 0.1 \times 2 \times 0.9 \times 0.1$ = 0.747 .
p_{12} = P(1 stays put & 1 comes across) + P(1 goes & 2 come)
         = $0.9 \times 2 \times 0.1 \times 0.9 + 0.1^3$ = 0.163 .
p_{13} = P(1 stays put and 2 come across) = 0.9×0.1^2 = 0.009 .

Note again that these probabilities sum to 1, although they do not form a Binomial probability distribution.

Now we have got the whole transition probability matrix, as states 2 and 3 are just the mirror images of states 1 and 0, looking at things from the point of view of the second container. So the complete matrix is:

$$
\begin{array}{c}
 \\ 0 \\ 1 \\ 2 \\ 3
\end{array}
\begin{array}{cccc}
0 & 1 & 2 & 3 \\
\left(\begin{array}{cccc}
0.729 & 0.243 & 0.027 & 0.001 \\
0.081 & 0.747 & 0.163 & 0.009 \\
0.009 & 0.163 & 0.747 & 0.081 \\
0.001 & 0.027 & 0.243 & 0.729
\end{array}\right)
\end{array}
$$

The stable probabilities for this matrix will tell us the proportion of time the gas is in the four different states. We could compute these in the usual way, but first let us consider the case when we change the probability of transfer from 0.1 to 0.5 . Using the same arguments as above, we get for P:

$$\begin{pmatrix} 0.125 & 0.375 & 0.375 & 0.125 \\ 0.125 & 0.375 & 0.375 & 0.125 \\ 0.125 & 0.375 & 0.375 & 0.125 \\ 0.125 & 0.375 & 0.375 & 0.125 \end{pmatrix}$$

The fact that each row is the same shows that the stable probabilities for this matrix can be read directly from any row. In fact, as you can check by multiplication, these are the same stable probabilities as for the previous example, with transfer probability of 0.1 . You might like to think a bit about why this should be so.

In general, with N molecules, the stable probability of i molecules being in the first container is Binomial:

P(i molecules in chamber 1)   = $^{N}C_{i}(\frac{1}{2})^{N}$.

13.6.4 SOCIAL PROCESSES

Suppose we assume 3 social classes (Lower, Middle and Upper), and that sons from one class move to another class according to the following one-generation transition probability matrix:

	Lower	Middle	Upper
Lower	4/5	1/5	0
Middle	1/4	5/8	1/8
Upper	0	1/4	3/4

If we start with the following distribution of the population into the classes: (0.7, 0.2, 0.1), then the distribution one generation further on is obtained by multiplying this by P to get: (0.61, 0.29, 0.1).

Ultimately, after many generation, the situation will stabilise into a distribution given by the stable probabilities = (5/11, 4/11, 2/11) = (0.4545, 0.3636, 0.1818).

Note that this model of social migration is very simplistic, and leaves of account all kinds of factors which could not be ignored in a more exact model.

212

13.7 Exercises

1. For the Markov chains defined by the following one-step transition probability matrices, draw diagrams to illustrate the possible transitions and describe each state as fully as possible:

a)
$$\begin{pmatrix} 1/2 & 0 & 1/4 & 0 & 0 & 1/4 \\ 0 & 0 & 0 & 1 & 0 & 0 \\ 1/3 & 0 & 1/3 & 0 & 0 & 1/3 \\ 0 & 0 & 0 & 0 & 1 & 0 \\ 0 & 1 & 0 & 0 & 0 & 0 \\ 0 & 0 & 0 & 0 & 0 & 1 \end{pmatrix}$$

b)
$$\begin{pmatrix} 0 & 1/2 & 1/2 & 0 & 0 \\ 1/3 & 2/3 & 0 & 0 & 0 \\ 0 & 0 & 1/8 & 7/8 & 0 \\ 0 & 1 & 0 & 0 & 0 \\ 1/4 & 1/4 & 0 & 0 & 1/2 \end{pmatrix}$$

c)
$$\begin{pmatrix} 0 & 1 & 0 & 0 \\ 1/2 & 0 & 1/2 & 0 \\ 0 & 1/2 & 0 & 1/2 \\ 0 & 0 & 1 & 0 \end{pmatrix}$$

d)
$$\begin{pmatrix} 0.3 & 0.1 & 0.5 & 0 & 0.1 \\ 0 & 1.0 & 0 & 0 & 0 \\ 0.2 & 0 & 0.8 & 0 & 0 \\ 0 & 0 & 0 & 0.7 & 0.3 \\ 0 & 0 & 0 & 0.6 & 0.4 \end{pmatrix}$$

2. A mouse is running about in a 6-roomed house, whose floorplan is as follows:

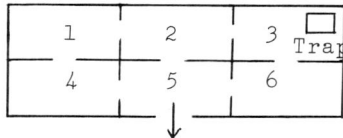

Figure 13.9

He moves from room to room, choosing the door by which he
leaves each room at random. If he is in room 5, he may
escape from the house via the front door. In room 3 there
is a mouse-trap, and if he is in room 3 he has a probability of $\frac{1}{2}$ of being trapped. Set up this situation as a
Markov chain, and write down the one-step transition
probability matrix. Describe the states of your chain.

3. Find the stable probabilities of the Markov chains with the
    following one-step transition probability matrices:

a)
$$\begin{pmatrix} 1/2 & 1/4 & 1/4 \\ 0 & 2/3 & 1/3 \\ 2/5 & 2/5 & 1/5 \end{pmatrix}$$

b)
$$\begin{pmatrix} 1/4 & 1/8 & 5/8 \\ 1/3 & 1/3 & 1/3 \\ 1/2 & 1/3 & 1/6 \end{pmatrix}$$

4. A model of the occurrence of geological strata assumes that
    there are three different rock types: sandstone, limestone
    and shale. Every vertical foot, the probability of
    transition from one rock type to another is given by:

	S	L	Sh
S	0.6	0.2	0.2
L	0.1	0.7	0.2
Sh	0.4	0.4	0.2

Find the relative proportions (in footage terms) of the
three different rock types. Change your model so that,
instead of using as steps each vertical foot of rock, you
use the points at which a change in rock type occurs (i.e.
the probability of transition to the same rock type is
zero). Find the relative proportions of the different
rock strata, irrespective of thickness.

5. An ideal gas consists of 4 molecules in 2 containers,
    connected by a thin tube. Every second, each molecule has
    a probability of $\frac{1}{4}$ of transferring to the other container.
    Set up a Markov chain model of this gas, and write down the
    one-step transition probability matrix. Find the stable
    probabilities.

6. Consider the following section of pseudo-Pascal, and
    assume that the three conditions cond1, cond2 and cond3 are

equally likely to be true or false each time they are
encountered:

```
repeat
  procA;
  if cond1
    then procB
    else
    begin
      procC;
      if cond2 then error;
    end;
until cond3;
```

Assume that if "error" is called the program terminates
with an error condition. Set up a Markov chain model of
this and calculate the expected number of times each
procedure will be called, as well as the probability of an
error termination.

7. The possible transitions of a Markov chain can be illustrated as follows:

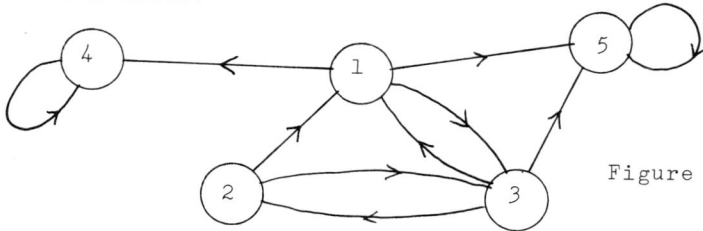

Figure 13.10

From every state, the transitions indicated all have equal
probabilities. Show that the fundamental matrix is:

$$N = \begin{pmatrix} 5/4 & 1/6 & 1/2 \\ 1/2 & 4/3 & 1 \\ 3/4 & 1/2 & 3/2 \end{pmatrix}$$

Hence find the matrix of absorption probabilities from
each non-absorbing state.

13.8 Computer Projects

1. Devise a program which will accept as input the one-step
transition probability matrix P for a Markov chain, and

output on a suitable graphics device a diagram showing the possible transitions between states.

2. Write a program to compute the stable probabilities for a regular Markov chain.

3. Write a program to compute the fundamental matrix for an absorbing chain, and also the matrix of ultimate absorption probabilities from each non-absorbing state.

4. Combine the previous programs into a general-purpose Markov chain analysis program. This should be able to diagram an input Markov chain, decide whether it is regular or absorbing, and produce the appropriate analysis.

CHAPTER 14

Queuing Theory

14.1 Introduction

Queuing theory is the study of the congestion caused by objects or individuals needing a common process or service, which takes a finite time and can only accommodate a finite number of items at once. Diagrammatically, we may represent this as follows:

Figure 14.1

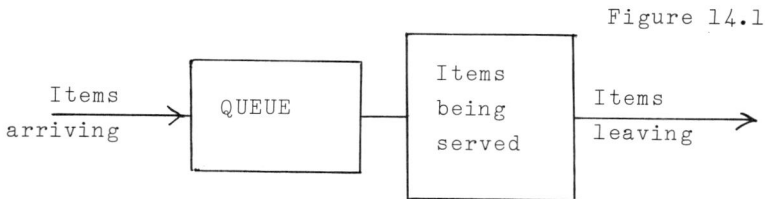

Queues arise in many applications, for example:
1. Post office counters.
2. Machines stopping and waiting for an attendant to reload or service them.
3. Cars waiting at traffic lights.
4. Jobs waiting to access a peripheral device on a computer.

From the above examples we can see that queues may be physical (people or objects waiting in a line to be served) or just an expression of a bottleneck in some kind of system. The same theory can be applied to either situation. Queues occur in many parts of computer systems, and an understanding of queuing theory will be very useful for those interested in systems analysis and design.

To study queues, we need a method for classifying them according to their properties. We shall consider three basic compnents of a queue:

1. The way in which items or "customers" arrive. This may be fixed (deterministic) or random. If it is random the times between successive arrivals may have various probability distributions, and these may be independent or dependent.

2. The rate at which customers are served, as well as the number of servers. Again, this may be deterministic or random.

3. The queue discipline, or rules for deciding which customer is served next. Quite commonly this is "fist in, first out" (FIFO), but sometimes we may have a stack, or "last in, first out" (LIFO) discipline. It is also possible to have a priority system, with a provision for high priority customers to be served ahead of low priority ones.

Obviously, it is possible for there to be interaction between the various components of a queue. For example, the service time may decrease if the queue length is long, or more servers may be brought into use. Or a long queue may discourage new users from joining it. The general queuing problem can thus become quite complex, and in this chapter we shall confine ourselves to simpler examples.

Before continuing with queues themselves, it is useful to discuss a very important mathematical model which is frequently used in situations where events occur randomly in time. This is the *Poisson Process*.

14.2 The Poisson Process

We assume that time is a continuous variable, and that events of some type (earthquakes, customers, bulbs failing) occur randomly in time.

Figure 14.2

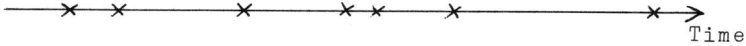

To set up the Poisson Process model, we need to make two assumptions:

1. If time is divided into very small intervals Δt, within each such interval the probability of an event occurring is $\lambda \Delta t$, where λ is a parameter which measures the "rate of occurrence" of events. Within the small time interval Δt the probability of more than one event occurring is so small that we may ignore it.

2. The numbers of events that occur in two disjoint time intervals are independent random variables.

From these two assumptions it is possible to work out the probability distributions of two interesting random variables:

    a) The number of events (r) that occur in a given time interval T.

    b) The length of time (X) between successive events.

To get P(Exactly r events in time T), divide the interval T into m very small intervals Δt, with $\Delta t = T/m$.

Figure 14.3

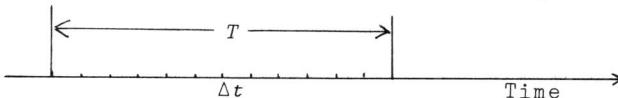

The probability of an event in each small interval $\Delta t = \lambda \Delta t$. The probability of more than one is taken to be 0. Thus we have a Binomial experiment, with m trials each with success probability $\lambda \Delta t$.

So $P(r$ events in time $T)$ $=$ ${}^m C_r (\lambda \Delta t)^r (1-\lambda \Delta t)^{m-r}$

$$= {}^m C_r (\lambda T/m)^r (1-\lambda T/m)^{m-r} .$$

If we let m tend to ∞, so that Δt tends to 0, then the Binomial probability distribution above tends to a Poisson distribution with the same mean.

$\qquad$ Mean $=$ $m \times \lambda \Delta t$ $=$ $m \lambda T/m$ $=$ λT .

Therefore, $P(r$ events in time $T)$

$$= \frac{e^{-\lambda T}(\lambda T)^r}{r!} \qquad \text{(Poisson distribution)} .$$

$\qquad$ $E(r)$ $=$ λT, and $\text{Var}(r)$ $=$ λT also.

$\qquad$ The random variable r is discrete, as it relates to the number of events in a given interval. The other random variable of interest is continuous - X, the time period between successive events. Assume that we measure time from the last event to have occurred.

$\qquad$ Let $R(x)$ $=$ P(No event before time x) .

$\qquad\qquad\qquad\qquad\qquad\qquad\qquad\qquad$ Figure 14.4

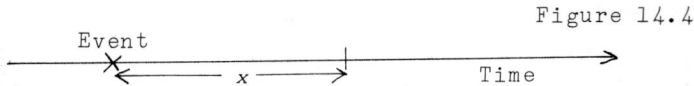

Event

$\xleftarrow{\quad\quad} x \xrightarrow{\quad\quad}$ $\qquad$ Time

Now $R(x)$ $=$ P(0 events in time x)

$\qquad\qquad$ $=$ $\dfrac{e^{-\lambda x}(\lambda x)^0}{0!}$ $\qquad$ (from the previous results)

$\qquad\qquad$ $=$ $e^{-\lambda x}$.

But $R(x)$ $=$ P(Inter-event time $X > x$)

$\qquad\qquad$ $=$ $1 - F(x)$, where $F()$ is the probability distrib-

$\qquad\qquad\qquad\qquad$ ution function for X.

Therefore, $F(x)$ $=$ $1 - R(x)$ $=$ $1 - e^{-\lambda x}$.

$\qquad$ This is the probability distribution function for the negative exponential distribution, and its probability density function is

$$f(x) = \lambda e^{-\lambda x} .$$

Thus we can see an interesting duality between the two random variables r and X:

| Number of events r in time T has a Poisson distribution. | <=> | Inter-event time X has a negative exponential distribution. |

It is also interesting to consider the slightly different random variable X^*, where X^* is the time from any arbitrary fixed point in time to the next event. Notice that in the above argument to derive the probability distribution for X, the fact that there was an event at time 0 was not used at all. Thus the same argument will give us exactly the same probability distribution for X^* as for X. This shows one slightly strange property of the Poisson Process - even if a long time has gone by since the last event, the probability distribution for the time to the next event is exactly the same as if an event had just happened.

Another way of looking at this is through the *survivor function*. Let

$$g(x)\Delta x = P(\text{Event in } (x, x+\Delta x)|\text{No event up to time } x)$$
$$= P(\text{Next event in } (x, x+\Delta x))/P(X > x)$$
$$= \lambda e^{-\lambda x}\Delta x / e^{-\lambda x} = \lambda\Delta x .$$

Thus $g(x) = \lambda$, which is independent of x. If we consider the event to be a "death", this implies that the chance of dying is the same regardless of how old you are. In practice, for humans and other organisms and systems, the survivor function $g(x)$ is age-dependent.

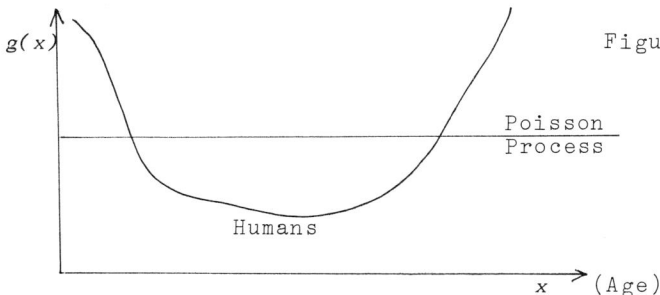

Figure 14.5

For humans, death is quite likely when very young or very old, but not so likely in the middle. It would not be reasonable, therefore, to model human life as a Poisson Process.

221

14.3 Properties of Queues

In order to classify queues, D.G. Kendall devised a three-part code, consisting of letter/letter/number. These specify:
1. The probability distribution for the arrival of customers.
2. The probability distribution for the service times.
3. The number of servers.

Some of the letter codes used are:

M - Negative exponential distribution, equivalent to a Poisson Process, as above. M comes from the name of A. Markov, who did a lot of the early work on queues.

D - Deterministic, or fixed time intervals.

G - General probability distribution.

Examples:

1. A hospital out-patient department, with a booking system, so that patients arrive at regular intervals, but the time taken to deal with each patient is Normally distributed. There are 6 doctors.

   This would be modelled as a D/G/6 queue.

2. A factory with 3 machine minders who reload machines when they stop - this takes 5 minutes to do. Machines stop randomly.

   This would be modelled as a queue with arrivals (machines stopping) as a Poisson Process, and deterministic service times - i.e. M/D/3.

When analysing a queue, we will usually be interested in its long-term behaviour. Obviously, the queue will start off sometime, but we shall ignore the initial period when it is settling down - this is equivalent to studying only the stable probabilities of a Markov chain. Things we may want to know about the queue in its steady state are:

1. Server utilisation - the proportion of time that the server is busy.
2. Average customer waiting time, between his arrival and the end of his service.
3. Average number of customers in the queue.
4. Average server busy period - the average time for which he is continuously busy.

(Note: throughout this study, the number in the queue will include the customer being served).

14.4 The M/M/1 Queue

This is one of the simplest types of queues, with both customer arrivals and service endings modelled as Poisson Processes. If X is the time between successive customers, then it has a negative exponential distribution with arrival rate λ per unit time:

$$f(x) = \lambda e^{-\lambda x} \quad \text{and} \quad E(X) = 1/\lambda \ .$$

There is a single server, and his service time Y is also negative exponential, with rate μ per unit time:

$$f(y) = \mu e^{-\mu y} \quad \text{and} \quad E(Y) = 1/\mu \ .$$

We are interested in the steady-state behaviour of this queue, so let us define:

$$p_i(t) = P(i \text{ customers in the queue at time } t).$$

Consider a small interval of time δt, so small that we can ignore the probabilities of more than one customer arriving or being served. Within this small time period, four possible things can happen:

Event	Probability
1. Another customer arrives	$\lambda \delta t$
2. A customer is served and leaves	$\mu \delta t$
3. One customer arrives and one leaves	$\lambda \mu \delta t^2$
4. No change	$(1-\lambda \delta t)(1-\mu \delta t)$

To get the probability of i customers in the queue at time $t+\delta t$, we can write:

$$p_i(t+\delta t) \;=\; p_{i-1}(t)\lambda\delta t \;+\; p_{i+1}(t)\mu\delta t$$
$$+\; p_i(t)(\lambda\mu\delta t^2 + (1-\lambda\delta t)(1-\mu\delta t)) \;.$$

Now, if the queue is in a stable state, the probabilities do not change with time, and $p_i(t+\delta t) = p_i(t)$.

So
$$p_i(t) \;=\; \lambda p_{i-1}(t)\delta t + \mu p_{i+1}(t)\delta t + p_i(t) + 2\lambda\mu p_i(t)\delta t^2$$
$$-\; (\lambda+\mu)p_i(t)\delta t$$

$$\Rightarrow \quad \lambda p_{i-1}(t)\delta t + \mu p_{i+1}(t)\delta t - (\lambda+\mu)p_i(t)\delta t + 2\lambda\mu p_i(t)\delta t^2 = 0$$

Ignoring the term in δt^2 and dropping the dependence on t,

$$\lambda p_{i-1} + \mu p_{i+1} - (\lambda+\mu)p_i \;=\; 0 \;.$$

In the special case where $i = 0$, the same kind of reasoning gives: $\quad \mu p_1 - \lambda p_0 \;=\; 0$.

From the above equation, we get:
$$p_1 \;=\; \frac{\lambda}{\mu}\, p_0 \;.$$

For $i = 1$: $\qquad \lambda p_0 + \mu p_2 - (\lambda+\mu)p_1 \;=\; 0 \;,$

from which we get: $\qquad p_2 \;=\; (\lambda/\mu)^2 p_0 \;.$

We can apply this method in sequence to obtain
$$p_3 \;=\; (\lambda/\mu)^3 p_0 \;,$$
$$p_i \;=\; (\lambda/\mu)^i p_0 \quad \text{etc.}$$

If we define $\rho = \lambda/\mu$, we can write
$$p_i \;=\; \rho^i p_0 \;.$$

To find the value of p_0, we make use of the fact that all the probabilities must sum to 1.
$$\sum_{i=0}^{\infty} \rho^i p_0 \;=\; 1 \;.$$

This is a geometric series, whose sum is $p_0/(1-\rho) \;=\; 1$. Thus $p_0 = 1-\rho$, and the probability distribution for queue length is
$$\underline{p_i \;=\; (1-\rho)\rho^i} \;.$$

This is actually the Geometric probability distribution (see chapter 4). Obviously, for this to give us a sensible probability distribution, we must have $\rho < 1$, i.e. $\lambda < \mu$, otherwise the queue will never have a steady state. From this

224

distribution we can compute most of the properties of the queue that we are interested in.

1. Server utilisation: Proportion of the time the server is not busy = P(0 customers in the queue) = p_0 = $1-\rho$.

   Therefore, server utilisation = $\rho = \lambda/\mu$.

2. Average number in queue; $= \sum_{i=0}^{\infty} i p_i = \sum_{i=0}^{\infty} i \rho^i (1-\rho)$

$$= (1-\rho)\rho \sum_{i=0}^{\infty} i \rho^{i-1} .$$

Now the summation gives the result $(1-\rho)^{-2}$ (check using the Binomial expansion), so the average number in the queue

$$= (1-\rho)\rho(1-\rho)^{-2} = \rho/(1-\rho) .$$

3. Average busy period:

   The server's time alternates from busy to idle. Over a long period T, the probability he is idle is p_0 (= $1-\rho$), and thus the total idle time is $T(1-\rho)$. Each idle period may be ended in a small time δt with probability $\lambda \delta t$ by the arrival of a new customer. Thus the length of an idle period is negatively exponentially distributed with mean $1/\lambda$. Therefore the total number of idle periods = $T(1-\rho)/(1/\lambda)$ = $T\lambda(1-\rho)$ = number of busy periods.

   The total busy time = $T\rho$, and thus the average busy period is of length

$$T\rho/T\lambda(1-\rho) = \frac{1}{\mu(1-\rho)} .$$

4. Average waiting time:

Time in queue before service

   The probability distribution for this is rather more complicated, but its mean can be shown to be $\rho/(\mu(1-\rho))$.

Example: A small computer system uses a single lineprinter. Jobs arrive randomly to be printed at an average rate of one per minute, and are printed with an average printing time of 0.8 minutes.

Modelling this as an M/M/1 queue:

$\lambda = 1.0$, $\mu = 1/0.8 = 1.25$. So $\rho = \lambda/\mu = 0.8$.
Thus, printer utilisation = ρ = 80% .
Average number of jobs waiting to be printed = $\rho/(1-\rho)$ = 4.
Average printer busy period = $1/(\mu(1-\rho))$ = 4 minutes.

Average waiting time $= \rho/(\mu(1-\rho)) = 3.2$ minutes.

The following program simulates an M/M/1 queue:

```
program qsim(input,output);
{ Program to simulate a simple queue. }
const maxq = 20;
      a = 519.7;
      b = 37.0;
type posint = 0..maxint;
var noinq,nits,ninit,i : posint;
    lambda,mu,dt,x : real;
    pinq : array[ 0..maxq ] of real;

function random(var x : real) : real;
{ Function to generate uniform pseudo-random nos. }
var z : real;
    i : posint;
begin
  z := a*x/b;
  i := trunc(z);
  z := z - i;
  x := z*b;
  random := z;
end;

procedure queue(lambda,mu,dt : real; var noinq : posint);
{ Procedure to simulate a single time step. }
begin
  if (random(x)<mu*dt) and (noinq>0) then noinq := noinq - 1;
  if random(x)<lambda*dt then noinq := noinq + 1;
end;

{ Main routine. }
begin
  for i := 0 to maxq do pinq[ i ] := 0.0;
  write('Input arrival rate (no./unit time): ');
  read(lambda); write(lambda:10:4); writeln;
  write('Input service rate (no./unit time): ');
  read(mu); write(mu:10:4); writeln;
  write('Input time step :');
  read(dt); write(dt:10:4); writeln;
  write('Input initial random no. : ');
  read(x); write(x:10:5); writeln;
  write('Input no. of time steps : ');
  read(nits); write(nits:6); writeln;
  write('Input initial no. of steps for stabilisation : ');
  read(ninit); write(ninit:6); writeln;
  noinq := 0;
  for i := 1 to ninit do queue(lambda,mu,dt,noinq);
  for i := 1 to nits do
  begin
    queue(lamda,mu,dt,noinq);
    if noinq <= maxq then pinq[ noinq ] := pinq[ noinq ] + 1
                     else pinq[ maxq ] := pinq[ maxq ] + 1;
end;
for i := 0 to maxq do pinq[ i ] := pinq[ i ]/nits;
```

```
   writeln('Queue probabilities for stable state - '); writeln;
   for i := 0 to maxq do
     writeln(i:5,pinq[ i ]:10:5);
end.
```

        Results from an example run of the program:

Input arrival rate (no./unit time):   1.0000
Input service rate (no./unit time):   1.2000
Input time step :   o.1000
Input initial random no. :   21.45700
Input no. of time steps :   5000
Input initial no. of steps for stabilisation :   1000
Queue probabilities for stable state -

        0   0.22020
        1   0.16360
        2   0.12860
        3   0.09580
        4   0.08780
        5   0.06500
        6   0.05780
        7   0.03840
        8   0.04200
        9   0.03440
       10   0.02700
       11   0.01320
       12   0.00660
       13   0.00840
       14   0.00800
       15   0.00280
       16   0.00040
       17   0.00000
       18   0.00000
       19   0.00000
       20   0.00000

14.5 Exercises

1. In the land of Quiva, earthquakes occur as a Poisson
    Process at a rate of one per month, and volcanoes erupt
    (as an independent Poisson Process) at the rate of two per
    year. Calculate the probability that in a 6-month period
    there will be four earthquakes and two volcanic eruptions.

2. In a certain small village murders usually take place at
    an average rate of one per year. Within one week three
    have occurred. Construct a hypothesis test, with a 1%
    significance level, to decide if this state of affairs is
    unusual.

3. A mini-computer system has a single disc drive, and the average disc access time is 2 milliseconds, negatively exponentially distributed. Jobs require the use of the disc randomly, at a rate of once every 3 milliseconds. Compute the properties of this queuing system.

4. An airport can accommodate 3 aircraft in 2 minutes for take-off or landing. If this rate is Poisson, what should be the mean time between arrivals (for landing or take-off) to ensure that the average waiting time is no more than 5 minutes ?

5. Consider an M/M/1 queue with discouragement, i.e. such that the arrival rate when i customers are in the queue
$$\lambda_i = \lambda/(i+1) \quad \text{for } i = 0,1,2, \ldots .$$
Assume that the service rate μ is constant, as before. Show that for the stable state, the probability of i customers in the queue is
$$p_i = \frac{e^{-\rho}\rho^i}{i!} \quad , \text{ where } \rho = \lambda/\mu .$$

14.6 Computer Projects

1. Adapt the simple queue simulation program to compute the server utilisation, mean queue length, average waiting time and average busy period.

2. Further adapt the program to consider the following variations:
   a) M/M/1 queue with discouragement (see exercise 5 above).
   b) M/M/n queue with n servers.

Suggested Further Reading for Section III

"Introduction to Probability Theory with Computing", by
J. Laurie Snell, Prentice-Hall, 1975.

"Queueing Theory in OR", by E. Page, Butterworths, 1972.

"Principles of Random Walk", by Frank Spitzer, Van Nostrand,
1964.

"Markov Chains", by D. Revuz, North-Holland, 1984.

"Applications of Queueing Theory", by G.F. Newell, Chapman and
Hall, 1982.

SECTION IV
Operations Research

CHAPTER 15

Linear Programming in Two Variables

15.1 Operations Research

Operations Research has been defined as the application of mathematics to management. In that sense, everything in this book is OR, since it is all concerned with using mathematical models to help us make decisions and predictions. In previous sections those mathematical models have been probabilistic in nature, as we have been dealing in situations involving a certain degree of uncertainty or randomness. In this section we shall consider OR problems which are deterministic, i.e. have no random element.

Before looking at some of these problems in detail, it is worth briefly discussing the general stategy for all OR work. Given a real problem to be solved in the real world, the approach adopted is as follows:

1. Formulate a mathematical model of the real situation, which expresses in as simple a form as possible the main features of the problem.
2. Express the problem in terms of the model, and solve mathematically in the model world.
3. Translate the model solution back into terms of the real problem, and evaluate critically its relevance.

In this general strategy, the first step is by far the

hardest. It requires experience, intuition and a certain amount of lateral thinking to see through the complex and confusing details of a messy problem to a useful mathematical model. Most useful of all is a model which we already know how to solve, but it may be hard to see that a particular model fits the new problem as well as other, apparently quite different, problems. In this book we cannot teach you how to do this, but by giving examples and exercises it is hoped that some of this skill will be developed.

Step 2 is what is usually considered to be OR - the manipulation of mathematical models to obtain solutions. Most of this section of the book will consist of developing algorithms and programs to do this in certain special cases. Step 3 is important, however, because we must take care to check our model result against the real world, in case our original model formulation omitted some factor of importance. It would be meaningless to tell an organisation to hire 3.814 new systems analysts ! Often we will find that the model needs to be re-formulated, and the whole process becomes iterative, as we seek out a model which gives an acceptable solution with a reasonable amount of effort.

Two important points should be borne in mind when solving OR problems:

1. KISS - "Keep it simple, stupid". Keep the model as simple as possible, consistent with the problem. A more complex model may be impossible to solve, give you no more information than a simple model, and leave you unable to see the wood for the trees.
2. Whenever possible, transform the problem into a form where it may be solved using a standard technique (as we did, for example, when doing non-linear regression in chapter 9). Remember Schagen's Law:
    "If you cannot solve a problem, always change it into a problem you can solve".

15.2 Introduction to Linear Programming

Linear Programming (LP) is a very powerful OR technique, and can be applied to a wide class of problems. We shall illustrate its use by means of an example.

Example: The Consolidated Grommets Problem

Consolidated Grommets Ltd. makes two products - flanged grommets at a net profit of 4p each, and knurled grommets at a net profit of 2p each. The maximum labour force they can use is 100 men. In one year a skilled grommet-maker can turn out 300,000 flanged grommets or 500,000 knurled grommets. However, the availability of raw materials (in particular high-impact grommetium) restricts the total manufacture of grommets of either type to 40 million in any one year. The wholesalers state that in the coming year they can only sell a maximum of 25 million flanged grommets and 30 million knurled grommets.

Bearing all these facts in mind, how many grommets of each type should be made in the coming year ?

To solve this kind of problem, we first need to formulate a mathematical model. To do this, we introduce *variables,* which represent the quantities about which a decision has to be made, or over which we have control. In this case we need to define two variables:

x_1 = Millions of flanged grommets made in the next year
x_2 = Millions of knurled grommets made in the next year
(Notice the choice of units to simplify the arithmetic).

In terms of these variables, we are clearly interested in maximising the total profit:

Z = $4x_1 + 2x_2$ (million pence).

Z is the *objective function,* whose value we wish to maximise. It is a linear function of our variables, so if we plot contours of constant Z on a graph whose axes are labelled x_1 and x_2, we get straight lines.

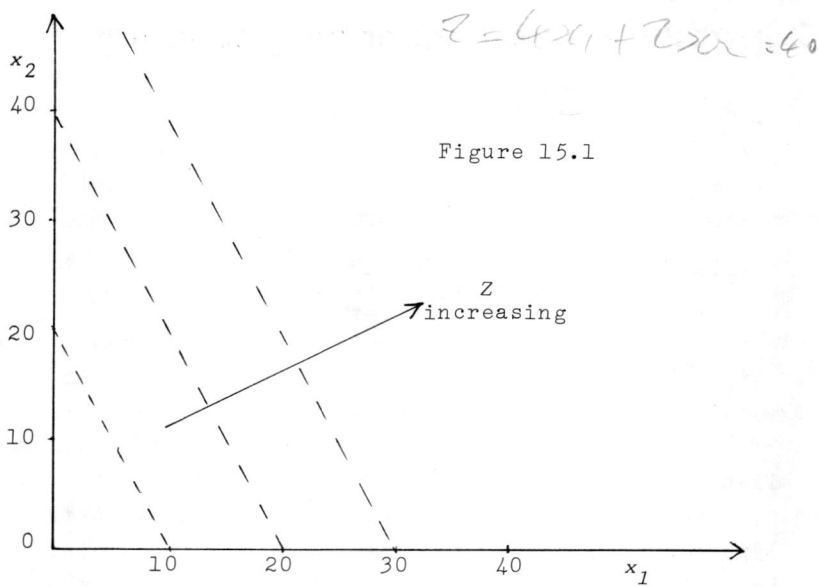

$Z = 4x_1 + 2x_2 = 40$

Figure 15.1

From this figure alone, it would appear that to maximise Z we just make x_1 and x_2 infinitely large - clearly this is impractical from the nature of the original problem. Our mathematical model is incomplete - it needs some *constraints*. These are mathematical representations of the physical restrictions in the original problem.

Constraint 1 is derived from the restriction on the available labour of 100 men.

To make x_1 million flanged grommets would take $x_1/0.3$ men, and to make x_2 million flanged grommets would take $x_2/0.5$ men.

So the total labour required is $x_1/0.3 + x_2/0.5$,

and the constraint is $x_1/0.3 + x_2/0.5 \leq 100$,

or $5x_1 + 3x_2 \leq 150$

(multiplying through by 1.5).

Constraint 2 is derived from the restriction on the total available materials, and is simply:

$$x_1 + x_2 \leq 40 .$$

Constraint 3 comes from the limit on the sales of flanged

grommets:
$$x_1 \leq 25 .$$

Constraint 4 comes from the limit on the sales of knurled grommets:
$$x_2 \leq 30 .$$

To finish off our mathematical model, we need to remember that it is impossible to make negative grommets, and this gives us the following *non-negativity restrictions*:
$$x_1 \geq 0 \text{ and } x_2 \geq 0 .$$

In total, our mathematical model can be expressed in the form of a linear programming problem:

Maximise $Z = 4x_1 + 2x_2$,
Subject to:

$$5x_1 + 3x_2 \leq 150$$
$$x_1 + x_2 \leq 40$$
$$x_1 \leq 25$$
$$x_2 \leq 30$$
$$x_1, x_2 \geq 0 .$$

This is called a linear programming problem because the objective function (Z) and the constraints are all linear functions of the variables. The solution of a general linear programming (LP) problem involves the use of the Simplex method, which is beyond the scope of this book. However, with just two variables, as here, it is possible to represent the LP problem graphically and thus solve it.

Each constraint will be represented by a line which divides the (x_1, x_2) plane into two regions - a region which satisfies the constraint and a region which does not. For example, to represent the constraint $5x_1 + 3x_2 \leq 150$, first draw the line $5x_1 + 3x_2 = 150$. We need two points to draw this line, so we consider what happens when $x_1 = 0$ (then $x_2 = 50$), and when $x_2 = 0$ (then $x_1 = 30$). This gives us a pair of points $(0,50)$ and $(30,0)$ which will define the correct line. Which side of the line satisfies the constraint ? Consider the point $x_1 = 0$, $x_2 = 0$: this does satisfy it (we say it is a *feasible* point) so that tells us which side of the line we have drawn represents the constraint.

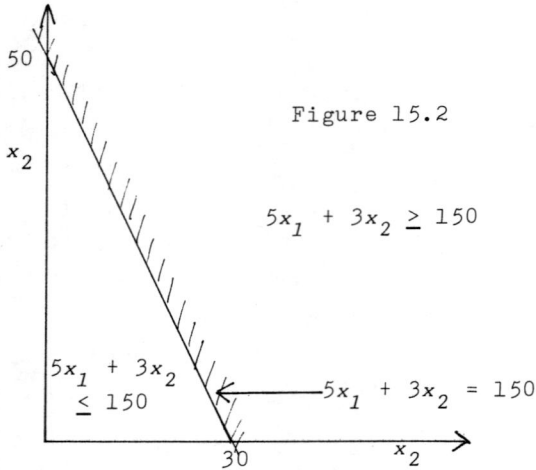

Figure 15.2

$$5x_1 + 3x_2 \geq 150$$

$5x_1 + 3x_2 \leq 150$

$5x_1 + 3x_2 = 150$

We shade the *infeasible* side of the line (the side for which the constraint is not satisfied) and thus the constraint is represented on the graph by all points on the line or on the unshaded side of the line. Putting all constraints together, we get:

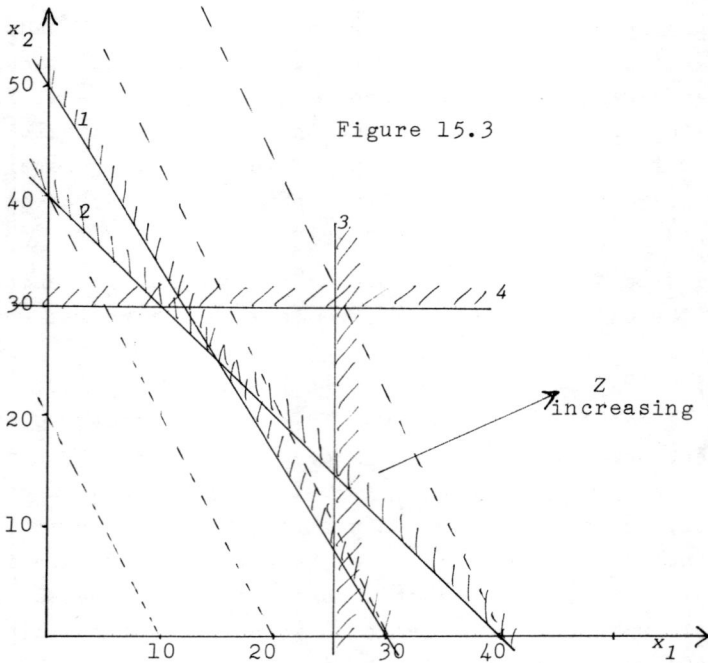

Figure 15.3

Z increasing

238

Remember also that $x_1 \geq 0$ and $x_2 \geq 0$, so that the set of all values of x_1 and x_2 that are possible is represented by the polygon bounded by the axes and the constraint lines. This is known as the *feasible region*, and we wish to find the point in there with the maximum value of Z. Consideration of the contours of Z shows fairly clearly that this will be at the corner where constraints 1 and 3 intersect. To find the exact values of x_1 and x_2 there, solve:

$$5x_1 + 3x_2 = 150$$
$$\text{and} \quad x_1 = 25 .$$

Therefore, $x_2 = 25/3 = 8.333$, and $Z = 116.667$ million pence.

This gives us the solution to our mathematical model problem of maximising Z subject to the given constraints. However, if we return to Consolidated Grommets Ltd. and tell them to make 25 million flanged grommets and 8.333 million knurled grommets, they will reply as follows:

"Our union will not accept your solution, since they insist (because the conditions in the knurling shop are more congenial than those in the flanging foundry) that at least as many knurled grommets as flanged grommets are made".

This is an extra condition which we did not take into account in the original model, but of course it invalidates the solution we have obtained. We need an extra constraint 5:

$$x_1 \leq x_2$$
$$\text{or} \quad x_1 - x_2 \leq 0 .$$

The new constraint, added to the existing constraints, gives a new feasible region which is a subset of the previous one, and a new maximum value of Z, which will be smaller than the previous one.

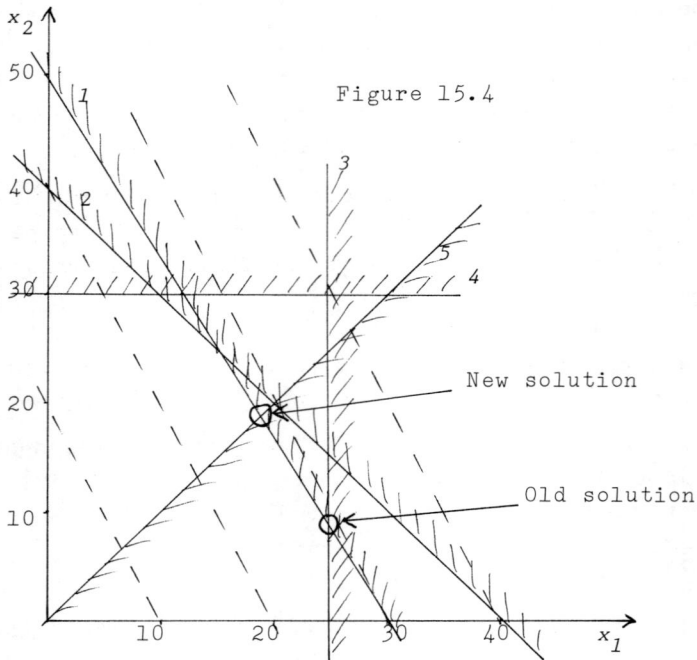

Figure 15.4

From this figure, it is clear that the new maximum Z value is found at the intersection of constraints 1 and 5.

$$5x_1 + 3x_2 = 150$$
$$x_1 - x_2 = 0 \; .$$

Therefore, $x_1 = x_2 = 18\frac{3}{4}$, and $Z = 112\frac{1}{2}$ million pence. This is the final solution which satisfies all the constraints.

Let us now briefly outline the steps involved in formulating and solving an LP model for a given problem.

1. Decide on the variables for the model - those quantities whose values we wish to determine, and over which we have control. (Keep the number of variables as small as possible, consistent with the problem).

2. Write down the objective function as a linear function of the variables, and decide whether to maximise or minimise.

3. Translate each constraint into a linear equality or inequality in the variables, making sure that nothing is omitted.

4. Solve the resulting LP problem (graphically, if there are only two variables).

5. Check the model solution against the problem, and see if it is sensible. If not, revise the model to take account of factors which have been left out, and re-solve.

From our example above, certain basic features of LP problems emerge which are true in general. Mathematically, we may define an LP problem to be of the form: find values of n variables $x_1, x_2, \ldots x_n$ which maximise (or minimise) a linear objective function of the variables, subject to a number of linear constraints (which may be equalities or inequalities) and to the condition that all the variables must be ≥ 0. With just two variables ($n = 2$) we can solve this graphically, and we define the feasible region to be the polygon bounded by the constraint lines and the axes within which lie all the values which satisfy the constraints.

For an LP problem the feasible region is always *convex*. This means that if any two points are in the feasible region, then all the points on the straight line between them will also lie in the feasible region.

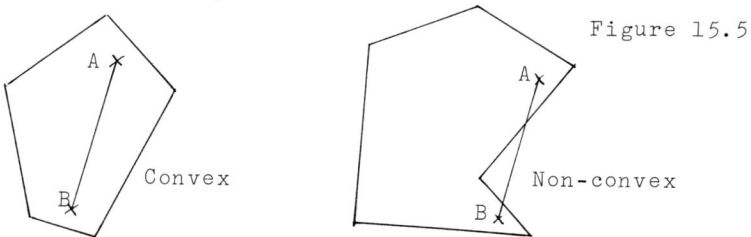

Figure 15.5

From the convex nature of the feasible region and the linearity of the objective function, it is fairly clear that the maximum (or minimum) value of Z will always be at a boundary point of the feasible region, in particular at one of the corners. This fact is very important in the general solution of LP problems in higher dimensions.

15.3 Special Cases in Linear Programming

Some special cases in LP can be illustrated using two-variable problems, and carry over into the many-variable situation also.

15.3.1 MULTIPLE OPTIMAL SOLUTIONS

If the contours of Z are parallel to one edge of the feasible region, then there may be multiple optimal solutions.

Example: Maximise $Z = x_1 + x_2$

Subject to:
$$x_1 + x_2 \leq 10$$
$$x_2 \leq 7$$
$$x_1 - x_2 \leq 5$$
$$x_1, x_2 \geq 0$$

Figure 15.6

x_2

A (3,7)

Z increasing

B $(7\frac{1}{2}, 2\frac{1}{2})$

x_1

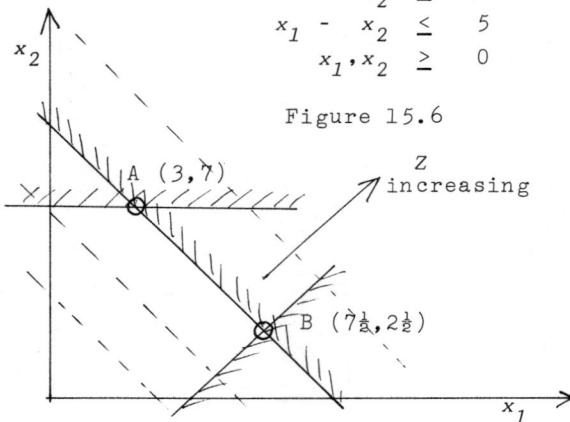

There are optimal solutions at both A and B, and anywhere on the line between them (an infinite number of points).

15.3.2 UNBOUNDED SOLUTIONS

In this case, there is no limit to Z, which may increase to ∞ (if maximising) or decrease to -∞ (if minimising).

Example: Maximise $Z = x_1 + x_2$

Subject to:
$$x_1 - x_2 \geq 0$$
$$x_1 - 2x_2 \geq -2$$
$$x_1, x_2 \geq 0$$

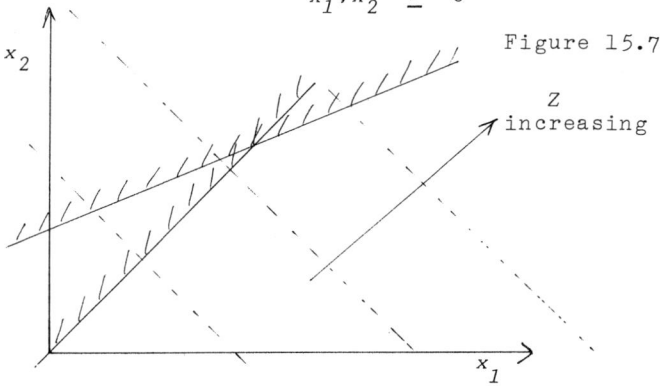

Figure 15.7

Z increasing

15.3.3 NO FEASIBLE SOLUTIONS

In this case the constraints are inconsistent, so that there is no feasible region at all.

Example: Maximise $Z = x_1 + 2x_2$

Subject to:
$$x_1 + x_2 \leq 2$$
$$2x_1 + x_2 \geq 5$$
$$x_1, x_2 \geq 0$$

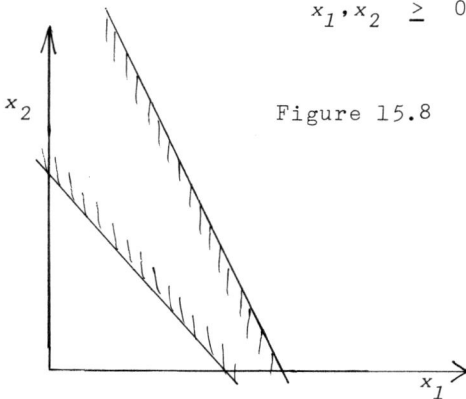

Figure 15.8

15.3.4 DEGENERACY

Normally in a two-variable LP, no more than two constraint lines or axes meet at a point. If more than two such lines meet at the same point, the LP is said to be degenerate there.

Example: Maximise $Z = 2x_1 + x_2$

Subject to:
$$x_1 + x_2 \leq 4$$
$$x_1 \leq 2$$
$$x_1 - x_2 \leq 0$$
$$x_1, x_2 \geq 0 \quad .$$

Figure 15.9

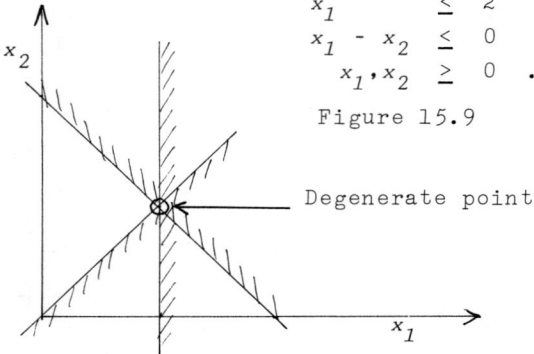

Degenerate point

15.4 Exercises

1. Mrs. Feasible, the OR man's wife, wants to provide a nutritious breakfast for her children. She can choose between two possible cereals, Snappies and Cracklies, and has to decide how to blend these cereals to give her children a nourishing breakfast at the lowest possible cost. She wants to make sure that each child has at least 200 milligrams of Vitamin A, 150 milligrams of Vitamin B, and 2 ozs. of roughage in his breakfast. The amounts of these ingredients in each cereal, and the costs per ounce, are given below:

Cereal	Vitamin A (mg/oz.)	Vitamin B (mg/oz.)	Roughage	Cost (per oz.)
Snappies	50	20	1/5	5p
Cracklies	30	40	2/3	7p

What is the cheapest way for Mrs. Feasible to feed her brood ?

2. General Solution has been given the task of forming an army consisting of infantry and cavalry. Cavalry are twice as effective as infantry, and in order to stand a reasonable chance in battle, the total effective strength of the army should be at least equal to 30,000 infantry. Horses eat twice as much as men, and the general will only have enough food to feed 60,000 men. To avoid brawls, he wants to make sure that the cavalry never outnumber the infantry by more than 10,000, and that the infantry never outnumber the cavalry by more than 20,000. An infantryman is paid one shilling a day, and a cavalryman two shillings (horses get paid nothing). What is the cheapest effective army the general can muster ?

3. Bert Basic makes home-made wine. In the coming year he is going to make elderberry and blackcurrant wine, but he only has 6 bottles. His wife insists that he should make at least half as much blackcurrant as elderberry, while his mother-in-law insists that the total alcohol content should be at least equivalent to five bottles of elder-berry (blackcurrant has three times the alcohol content of elderberry). Elderberry is twice as sweet as blackcurrant, and Bert wants to be sure that the total sweetness is at least the same as five bottles of blackcurrant. What is the smallest number of bottles of blackcurrant that Bert need make ? What is the largest number of bottles of elderberry that he can make ?

4. For the following LP problems, plot diagrams showing the constraints, feasible region and objective function contours. Describe any special features of each problem.

a) Minimise $Z = 2x_1 + 3x_2$

Subject to:
$$x_1 + 3x_2 \leq 10$$
$$4x_1 + 6x_2 \geq 10$$
$$2x_1 - x_2 \geq 0$$
$$x_1, x_2 \geq 0$$

b) Maximise $Z = x_1 + x_2$

Subject to:
$$2x_1 - x_2 \leq 2$$
$$3x_1 + 4x_2 \geq 10$$
$$x_1, x_2 \geq 0$$

c) Maximise $Z = 5x_1 + 3x_2$

Subject to:
$$2x_1 + 3x_2 \geq 12$$
$$5x_1 + 3x_2 \leq 21$$
$$x_1 - x_2 \geq 2$$
$$x_1, x_2 \geq 0$$

5. The Computer Studies Department at a certain university runs two courses - Computer Processing (CP) and Data Studies (DS). Each year they must decide how many new students to admit to each course. The maximum intake for the whole department is fixed at 50. The available computer resources for first year students amounts to 6,000 units per week - each CP student requires 100 units per week and a DS student needs 150 per week. The probability of a CP student passing his first year exams and proceeding to the second year is 75%, while for DS students it is 80%. The department wishes the expected number of second year students to be at least 20. It has also been decreed that in the first year the ratio of CP to DS students should be at least 1.5 to 1. The "credit" the department gets for each CP student is 100, while for each DS student it is 120. Formulate this as an LP problem and find the optimal intake to maximise the total credit.

15.5 Computer Projects

1. Design a computer program to output on a suitable graphical device a plot of the feasible region of an LP problem, together with contours of its objective function.

2. Add to the above program a routine which will determine the optimal solution. This can be achieved by finding all the corners of the feasible region and choosing the one with the highest (or lowest) value of Z.

CHAPTER 16

The Simple Transportation Problem

16.1 Introduction

The Simple Transportation problem is in fact a special case of the general LP problem, but the way in which it is solved is quite instructive and interesting. To start with, let us give an example of this type of problem.

Example: Every week Consolidated Grommets Ltd. make a total of 50,000 grommets in its three factories, distributed as follows:

        Factory 1:   20,000
        Factory 2:   15,000
        Factory 3:   15,000 .

These grommets need to be transported to four wholesalers' depots, and the weekly requirements at each depot are as follows:

        Depot 1:   12,000
        Depot 2:   9,000
        Depot 3:   15,000
        Depot 4:   14,000 .

The costs of transportation from each factory to each depot have been computed, and are set out below, in pence per grommet transported:

	To				
		D_1	D_2	D_3	D_4
	F_1	4	6	5	8
From	F_2	7	5	6	5
	F_3	5	6	8	6

The problem clearly is to arrange the transportation of grommets from factories to depots in such a way as to minimise the total transportation cost. Before discussing the solution, let us first write down the problem in its most general form. Assume we have m "sources" of goods, each with an available quantity a_i, for $i = 1, ..m$. There are also n "destinations", each requiring an amount b_j, for $j = 1, ..n$. Let us further assume that

$$\sum_{i=1}^{m} a_i = \sum_{j=1}^{n} b_j$$

(We shall discuss later what to do if this is not true). The cost of transporting one unit from source i to destination j is c_{ij}, and we wish to find the quantities $\{x_{ij}\}$, where x_{ij} is the amount to be sent from source i to destination j.

Problem: Minimise $Z = \sum_{i=1}^{m} \sum_{j=1}^{n} c_{ij} x_{ij}$

Subject to: $\sum_{j=1}^{n} x_{ij} = a_i$ for $i = 1, ..m$

(i.e. amount sent from each source equals the amount available at that source)

and $\sum_{i=1}^{m} x_{ij} = b_j$ for $j = 1, ..n$

(i.e. amount received at each destination equals the amount required at that destination)

and $x_{ij} \geq 0$ for all i and j.

This is obviously an LP problem with mn variables and $m+n$ constraints (in our example, 12 variables and 7 constraints). However, it is possible to solve this in a simpler fashion than the general LP problem.

16.2 Solution Method

The first step is to introduce the concept of a *basic solution*. This is a set of $\{x_{ij}\}$ values which satisfy the constraints, and in which no more than $m+n-1$ variables are greater than 0. For example, in the above problem with $m = 3$ and $n = 4$, a basic solution will contain 6 *basic* variables, with values ≥ 0, and the remaining 6 *non-basic* variables will be set equal to 0. An example of such a basic solution is:

		D_1	D_2	D_3	D_4	a_i
	F_1	12	8			20
From	F_2		1	14		15
	F_3			1	14	15
	b_j	12	9	15	14	

To (above table columns D_1–D_4)

Thus only 6 of the 12 possible routes are actually used - these correspond to the 6 basic variables in this solution. There are clearly a large number of such basic solutions, depending on which variables are chosen to be basic and which non-basic.

An important result is that an optimal solution to the Simple Transportation problem is always to be found at a basic solution. This cuts down the number of possibilities we need to explore, since we only need to consider basic solutions. This is the same as the case in the two-variable LP problem (see chapter 15) in which the optimal solution always occurs at a corner of the feasible region. So a basic solution is equivalent to a corner of the feasible region (although to draw a picture of the feasible region we should need 12-dimensional graph paper).

We want therefore to find the basic solution which has the lowest value of the total transportation cost Z. Given

that we have a basic solution, how do we tell whether or not
it is an optimal solution ? If it is not, how do we transform
it into an optimal basic solution ? To answer these questions
we use the idea of *shadow costs*. Given a basic solution, with
$m+n-1$ values ≥ 0 and the rest set equal to 0, let:

u_i = Apparent or shadow cost (according to the current basic
solution) of taking one unit of goods from source i,

v_j = Apparent or shadow cost of delivering one unit to
destination j.

We can compute the shadow costs relating to the current
basic solution bey setting $u_i + v_j = c_{ij}$ for each basic
variable x_{ij}. For the example basic solution above, we get:

$$u_1 + v_1 = c_{11} = 4$$
$$u_1 + v_2 = c_{12} = 6$$
$$u_2 + v_2 = c_{22} = 5$$
$$u_2 + v_3 = c_{23} = 6$$
$$u_3 + v_3 = c_{33} = 8$$
$$u_3 + v_4 = c_{34} = 6 \ .$$

Notice that we have 6 equations and 7 unknowns, so to solve
we must arbitrarily assign a value to one of the unknowns.
Setting $u_1 = 0$ and solving for the others, we have:

$$u_1 = 0 \qquad v_1 = 4$$
$$u_2 = -1 \qquad v_2 = 6$$
$$u_3 = 1 \qquad v_3 = 7$$
$$v_4 = 5 \ .$$

Using these values, we can calculate for each non-basic
route (ij) the apparent cost $u_i + v_j$. If this apparent cost
is greater than the true cost c_{ij}, then it will pay us to
bring this route into operation and make the corresponding
variable basic. This would imply that the current basic
solution is not the optimal solution. On the other hand, if
$u_i + v_j \leq c_{ij}$ for all the non-basic variables, then the
current solution is optimal.

So, for all the non-basic variables, we shall compute
$u_i + v_j - c_{ij}$. If these values are all ≤ 0, this is an optimal
solution. Otherwise, we shall choose the most positive of
these values to indicate which non-basic variable should

become basic in the new solution.

Let us return to our example problem, and indicate the current basic solution with the shadow costs included.

	D_1	D_2	D_3	D_4	u_i
F_1	4 / (12)	6 / (8)	5 / +2	8 / -3	0
F_2	7 / -4	5 / (1)	6 / (14)	5 / -1	-1 ($Z = 277$)
F_3	5 / 0	6 / +1	8 / (1)	6 / (14)	1
v_j	4	6	7	5	

The notation we shall use for these tables allows all the information to be displayed in a compact form:

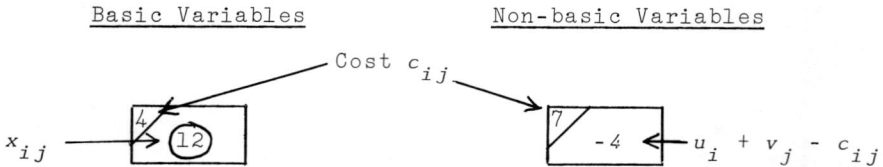

Basic Variables   Non-basic Variables

Cost c_{ij}

x_{ij} → 4 / (12)   7 / -4 ← $u_i + v_j - c_{ij}$

So the above solution is not optimal, because of the positive values of $u_i + v_j - c_{ij}$. We choose the non-basic variable with the highest positive value to become basic - this is x_{13}, which will become basic with value θ, say. To determine what θ should be and what effect it will have on the other basic variables, we use the concept of a *loop*.

A loop is a path from one cell to another in the table, such that steps are taken along rows or columns, no cell is repeated except that the first cell is the same as the last cell, and each step is at right angles to the previous step.

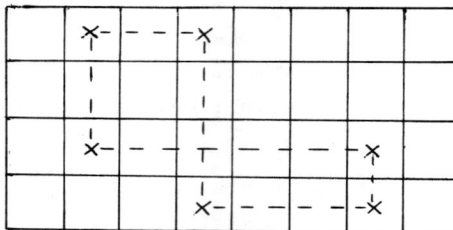

Figure 16.1

There are two rules relating the concept of a loop to a basic solution of a Simple Transportation problem:

1. A loop cannot be formed using just basic cells.
2. Any non-basic cell, together with some basic cells, can form a loop.

We use the latter type of loop to change the values of the basic variables and eliminate one of them from the solution. This is done by adding and subtracting the value of θ round the loop, so that the row and column totals are unchanged:

	D_1	D_2	D_3	D_4
F_1	12	8-θ	θ	
F_2		1+θ	14-θ	
F_3			1	14

We now choose θ so that one basic variable becomes 0 (i.e. non-basic) and the rest remain ≥ 0. To do this, choose θ equal to the smallest value from which θ is subtracted in the loop - in this case, θ = 8. Thus we obtain a new basic solution with 6 non-zero values, and with a lower value of the total cost Z:

	D_1	D_2	D_3	D_4	u_i
F_1	[4] (12) θ	[6] -2	[5] (8) +θ	[8] -5	0
F_2	[7] -2	[5] (9)	[6] (6)	[5] -1	1 (Z = 261)
F_3	[5] +2 θ	[6] +1	[8] (1) -θ	[6] (14)	3
v_j	4	4	5	3	

Again, we compute the u_i and v_j values (this can easily be done by inspection of the table, without even writing down the equations) and fill in $u_i + v_j - c_{ij}$ for the non-basic variables. Again, we have a non-optimal solution, so we must make x_{31} basic in the next iteration. The loop is shown in the above table, and the necessary value of θ is 1. Notice that we can predict the reduction in Z for the next table - it will be equal to θ times the value of $u_i + v_j - c_{ij}$ for the new basic cell = 1×2 = 2. So we expect the next value of Z to be 259. The new table is:

253

	D_1	D_2	D_3	D_4	u_i
F_1	4 ⑪−θ	6 −2	5 ⑨+θ	8 −3	0
F_2	7 −2	5 ⑨	6 ⑥−θ	5 +1 θ	1 (Z = 259)
F_3	5 ①+θ	6 −1	8 −2	6 ⑭−θ	1
v_j	4	4	5	5	θ = 6 .

	D_1	D_2	D_3	D_4	u_i
F_1	4 ⑤	6 −1	5 ⑮	8 −3	0
F_2	7 −3	5 ⑨	6 −1	5 ⑥	0 (Z = 253)
F_3	5 ⑦	6 0	8 −2	6 ⑧	1
v_j	4	5	5	5	

This is an optimal solution to our problem, since all the values $u_i + v_j - c_{ij} \leq 0$ for the non-basic variables. You can see it took us just three iterations to get from the initial solution to the final, optimal solution. In outline, therefore, the Simple Transportation problem is solved as follows:

1. Set up an initial basic solution to the problem.
2. Compute u_i and v_j values and hence $u_i + v_j - c_{ij}$ for each non-basic cell. If these are all $\leq$ 0, the solution is optimal.
3. If it is not optimal, find the non-basic cell with the largest such positive value and make it basic with value θ.
4. Form a loop with this cell and existing basic cells, subtracting and adding θ. Hence find θ. Produce a new basic solution, and go to step 2.

16.3 Setting up the Initial Basic Solution

We have discussed in detail all the above steps, except the first. There are various methods for finding an initial basic solution, and they all follow the same fundamental pattern. We wish to make exactly $m+n-1$ cells basic and the rest non-basic (with value zero), so that all $m+n$ row and column constraints are satisfied. All methods for doing this involve, at each stage, picking a cell and making it basic with a value which completely satisfies *either* a row constraint *or* a column constraint, but *not both*, except for the last cell which does satisfy both. The main variation between the different methods is in the choice of the basic cell at each stage.

16.3.1 NORTH-WEST CORNER

Start at row 1, column 1 (the so-called north-west corner), and move to the next row when the row constraint is satisfied, and to the next column when the column constraint is satisfied. Thus the basic cells tend to lie close to the main diagonal of the table. This method is simple to operate, but since it takes no account whatever of the costs of the cells it may result in an initial solution far from the optimal solution. The initial solution for the previous example was generated by this method - the initial value of Z was 277.

16.3.2 COLUMN MINIMA

In this case, we start in column 1 and move on to the next column as soon as each column constraint is totally satisfied. Within a column, we choose the available cell with the lowest cost (cells whose row constraints are already satisfied are not available). For example:

Total cost $Z = 261$.

255

This method is identical to the previous one, except that it proceeds row by row instead of column by column.

	D_1	D_2	D_3	D_4	a_i		
F_1	4 ⓬	6	5 ⑧	8	~~20~~	~~8~~	✓
F_2	7	5 ⑨	6	5 ⑥	~~15~~	~~6~~	✓
F_3	5	6	8 ⑦	6 ⑧	~~15~~	7	
b_j	~~12~~	9	~~15~~	~~14~~			
	✓	✓	7	~~8~~ ✓			

Total cost Z = 267.

16.3.4 MATRIX MINIMA

In this case, we just pick each time the available cell with the lowest cost in the whole matrix, irrespective of row or column. If two cells have the same lowest cost, then we just make an arbitrary choice. Applying this method to our example problem gives the same initial solution as for the row minima method above, although this will not be true in general.

In this example problem, the methods which make use of the costs of the routes have given a lower initial total cost than the North-West corner method, which does not. This will be the case more often than not, and will usually mean that fewer iterations need be performed to find the optimal solution.

16.4 Special Conditions in Simple Transportation

16.4.1 MULTIPLE OPTIMAL SOLUTIONS

It may be the case, when we have an optimal solution,

that one or more of the values of $u_i + v_j - c_{ij}$ for the non-basic variables may exactly equal 0, although none are > 0. Our example problem has an optimal solution for which this is true. The variable x_{32} has $u_i + v_j - c_{ij} = 0$ in the optimal table, and this means that making x_{32} basic would in fact leave the value of Z unaltered. In other words, there is an alternative optimal solution, obtained by making x_{32} basic.

	D_1	D_2	D_3	D_4	u_i
F_1	4 (5)	6 -1	5 (14) ↗15	8 -3	0
F_2	7 -3	5 (9)$_\theta$	6 -1	5 (6)$_{+\theta}$	0 (Z = 253)
F_3	5 (7)	6 0 $_\theta$	8 -2	6 (8)$_{-\theta}$	1
v_j	4	5	5	5	$\theta = 8$

Alternative solution:

	D_1	D_2	D_3	D_4	u_i
F_1	4 (5)	6 -1	5 (14) ↗15	8 -3	0
F_2	7 -3	5 (1)	6 -1	5 (14)	0 (Z = 253)
F_3	5 (7)	6 (8)	8 -2	6 0	1
v_j	4	5	5	5	

16.4.2 DEGENERACY

A basic solution is degenerate if one or more of the basic variables is actually equal to zero. If this occurs, we must be careful not to confuse it with a non-basic variable, since we need $m+n-1$ basic variables for our solution method to work. We must distinguish between "basic zeroes" and non-basic variables which are zero by definition. Note that degeneracy can occur at any stage of the solution, unlike the case of multiple optimal solutions which can only occur in the final, optimal table. Degeneracy can appear in two ways during the solution algorithm:

1. In setting up the initial basic solution. If we select a basic variable (other than the last) which simultaneously

satisfies both a row and a column constraint, then we must "cross off" one or the other constraint, and say that 0 remains to be satisfied at the other. This means that we must put in a basic variable at level 0 to satisfy that constraint.

2. During an iteration of the solution method. If the chosen value of θ makes more than one basic variable become equal to 0, then we must choose just one to be non-basic, and the rest to be degenerate basic variables.

Example: (Revision of previous example - initial solution by North-West corner method).

	D_1	D_2	D_3	D_4	u_i
F_1	4 (12)	6 (8)$_\theta$	5 (0)$_{+\theta}$	8 −4	0
F_2	7 −2	5 +2$_\theta$	6 (15)$_{-\theta}$	5 (0)	1 ($Z = 276$)
F_3	5 +1	6 +2	8 −1	6 (15)	2
v_j	4	6	5	4	$\theta = 8$

Note the two degenerate zero values included in the 6 basic variables. One of these degenerate variables disappears in the next table.

	D_1	D_2	D_3	D_4	u_i
F_1	4 (12)$_{-\theta}$	6 −2	5 (8)$_{+\theta}$	8 −4	0
F_2	7 −2	5 (8)	6 (7)$_{-\theta}$	5 (0)$_{+\theta}$	1 ($Z = 260$)
F_3	5 +1$_\theta$	6 0	8 −1	6 (15)$_{-\theta}$	2
v_j	4	4	5	4	$\theta = 7$

(The other degenerate variable disappears).

	D_1	D_2	D_3	D_4	u_i
F_1	4 (5)	6 −1	5 (15)	8 −3	0
F_2	7 −3	5 (8)	6 −1	5 (7)	0 ($Z = 253$)
F_3	5 (7)	6 0	8 −2	6 (8)	1
v_j	4	5	5	5	

This is an optimal solution, with no degenerate variables.
However, there is an alternative optimal solution, which is
degenerate.

16.4.3 TOTAL AVAILABLE NOT EQUAL TO TOTAL REQUIRED

If $\sum_{i=1}^{m} a_i \neq \sum_{j=1}^{n} b_j$, then we need to make some
changes before we can solve the problem. There are two cases
to consider:

1. $\sum_{i=1}^{m} a_i > \sum_{j=1}^{n} b_j$.

There is no real difficulty in solving this problem, as
there is more available at the sources than is required at the
destinations. We need to include a "dummy" destination, with
transportation costs equal to 0, to represent goods left behind
at the sources. This extra destination should have requirements equal to $\sum a_i - \sum b_j$, and with this addition the problem
can be solved as before.

2. $\sum_{i}^{m} a_i < \sum_{j}^{n} b_j$.

This problem cannot be solved as it stands without
further information, as there will be shortfalls at some of the
destinations. If we know the costs of such shortfalls, they
can be included as the transportation costs of a "dummy" source,
and the problem solved as before.

16.5 Computer Implementation

Although this method of solution is very simple to carry
out by hand, there are some interesting features of it when a
computer implementation is required. A very simple diagram of
the algorithm might look as follows:

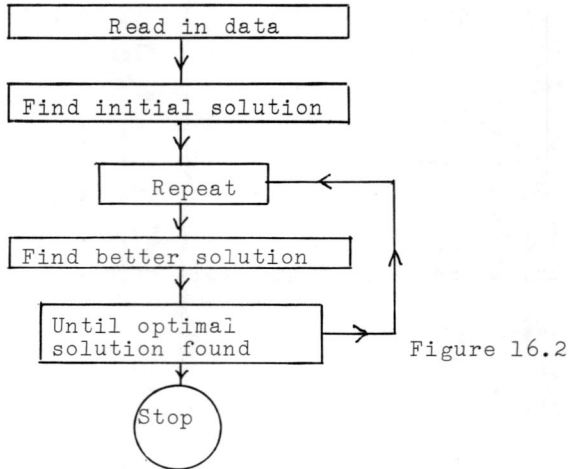

Figure 16.2

If we study the second box above ("Find initial
solution") and decide to implement the North-West corner
method, the following procedure might be developed:

```
program simtrans(input,output);
const maxvars = 19;
type posint = 0..maxint;
     coordtype = array[ 0..maxvars,1..2 ] of posint;
     intarr = array[ 0..maxvars ] of posint;
var ixy : coordtype;
    pred,a,b,val : intarr;

procedure nwcorn(var ixy : coordtype; var val,a,b : intarr;
                 nrows,ncols : posint);
{ Procedure to find initial basic feasible solution by
  NW corner method. }
var i,j,k : posint;
begin
  i := 1; j := 1; k := 0;
  repeat
    k := k + 1;
    if a[ i ] <= b[ j ]
    then begin
      val[ k ] := a[ i ]; a[ i ] := 0; b[ j ] := b[ j ] -
                                                 val[ k ];
      ixy[ k,1 ] := i; ixy[ k,2 ] := j; i := i + 1;
    end
    else begin
      val[ k ] := b[ j ]; b[ j ] := 0; a[ i ] := a[ i ] -
                                                 val[ k ];
      ixy[ k,1 ] := i; ixy[ k,2 ] := j; j := j + 1;
    end;
  until k = nrows + ncols - 1;
end;
```

The positions and values of the basic variables found in

this way are stored in the following data structures:

val $\lceil$ k $\rceil$ = value of the kth basic variable,

ixy $\lceil$ k,1 $\rceil$ = row number of the kth basic variable,

and ixy $\lceil$ k,2 $\rceil$ = column number of the kth basic variable.
Note that in the procedure the input values of a and b, the
available and required values, are destroyed.

Turning now to the problem of finding a better basic
solution, we may expand that part of the algorithm as follows:

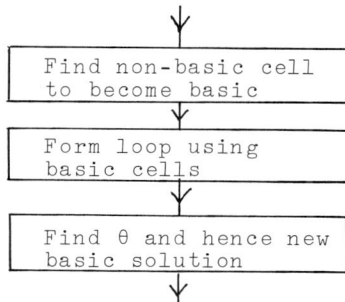

```
      +--------------------+
      | Find non-basic cell |
      | to become basic    |        Figure 16.3
      +--------------------+
      | Form loop using    |
      | basic cells        |
      +--------------------+
      | Find θ and hence new|
      | basic solution     |
      +--------------------+
```

Let us investigate implementing the middle box, finding
a loop. This is not as easy to program as it usually is to do
by eye.

Given: Positions of all basic cells, plus the position of the
non-basic cell to become basic.

Find: A loop linking the non-basic cell and some of the
basic cells.

To keep track of the basic cells, let us number them
1 ... $m+n-1$, and number the non-basic cell 0. Note that the
values of the basic variables are irrelevant to this part of
the algorithm.

Example: A 4×5 problem with 8 basic variables marked, plus a
non-basic cell (number 0):

```
+----+----+----+----+----+
| ①  |    |    | ②  |    |
+----+----+----+----+----+
|    |    | ③  | 0  |    |
+----+----+----+----+----+
|    | ④  | ⑤  |    | ⑥  |
+----+----+----+----+----+
| ⑦  |    | ⑧  |    |    |
+----+----+----+----+----+
```

To find a loop we shall adopt the following procedure:

1. Find all the basic cells on the same row as cell 0.
2. For a given cell, find all the basic cells on the same row (or column).
3. Alternate searches from rows to columns or columns to rows, unitl cell 0 is reached again.

Graphically, we can describe this procedure as a tree search:

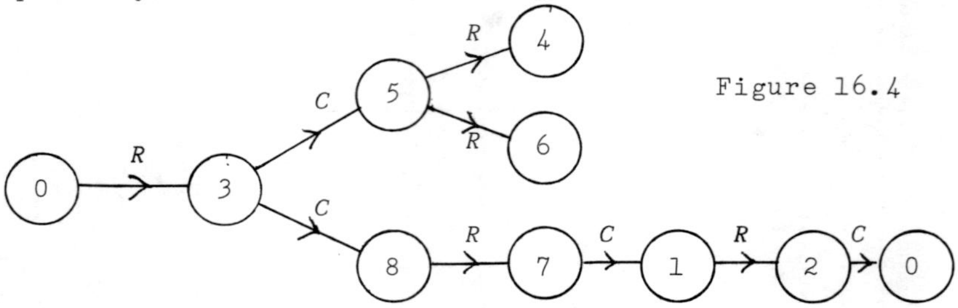

Figure 16.4

(R means two cells are on the same row, and C that they are on the same column).

If we define for each basic cell i, a variable

pred$[\,i\,]$ = cell before i in the tree,

then to find the loop we just go from 0 to pred$[\,0\,]$, and then back to its predecessor, and so on unitl reaching 0 again (thus we go from right to left through the tree diagram above). For our example, the loop becomes $0 \to 2 \to 1 \to 7 \to 8 \to 3 \to 0$.

The simplest way of programming a tree search of this type is via a *recursive* procedure (one which calls itself). A procedure treebuild has been written which carries out the operation of building the tree and hence finding the loop recursively. Apart from previously defined variables, it uses the following parameters:

        node :   Number of the input node (basic cell).
        colswitch :   = 1 if the search is along rows,
                      = 2 if it is along columns.
        npts :   Total number of basic cells = nrows + ncols
                                                    - 1.

```
procedure treebuild(var ixy : coordtype; var pred : intarr;
                    var node,colswitch,npts : posint);
{ Procedure to find loop joining basic cells. }
var i,newswitch,newnode : posint;
begin
  if ixy[node,colswitch] = ixy[0,colswitch]
  then pred[0] := node
  else begin
    for i := 1 to npts do
    begin
      if (ixy[node,colswitch] = ixy[i,colswitch]) and
         (i <> node)
      then begin
        pred[i] := node; newswitch := 3 - colswitch;
        newnode := i;
        treebuild(ixy,pred,newnode,newswitch,npts);
      end;
    end;
  end;
end;

{ Section of main routine calling treebuild. }
  for i := 1 to npts do
    if ixy[i,1] = ixy[0,1]
    then begin
      node := i; colswitch := 2; pred[node] := 0;
      treebuild(ixy,pred,node,colswitch,npts);
    end;
```

16.6 Exercises

1. A motor manufacturer needs to despatch cars of a certain
    model to 4 showrooms. The numbers required are 1,6,2 and
    6, and there are 5 cars available at each of 3 factories.
    The times in hours to drive from each factory to each
    showroom are given below.

		To			
		S_1	S_2	S_3	S_4
	F_1	5	4	3	2
From	F_2	10	8	4	7
	F_3	9	9	8	4

    Find a solution which minimises the total driving time.

2. 64 poems are required for a special royal celebration,
    being 16 odes, 20 elegies and 28 panegyrics. 4 poets have

been commissioned, P_1, P_2, P_3 and P_4, who have sufficient inspiration to write 12,17,21 and 14 poems respectively. The number of days required by each poet to write each style of poem are set out below.

	Ode	Elegy	Panegyric
P_1	3	7	8
P_2	3	6	5
P_3	7	2	9
P_4	10	9	4

Solve this problem to minimise the total number of poet-days. How many different optimal solutions are there?

3. Find an optimal solution of the following Simple Transportation Problem:

		Destinations					
		A	B	C	D	E	Available
	1:	20	47	17	41	62	8
Sources	2:	74	13	52	40	32	6
	3:	60	31	51	71	68	4
	4:	39	41	37	21	38	7
Required:		3	6	5	6	5	

? 4. Find an optimal solution of the following Simple Transportation Problem:

evry.

		Destinations				
		A	B	C	D	Available
	1:	10	25	40	50	6
	2:	20	15	35	30	5
Sources	3:	35	30	30	55	8
	4:	15	20	20	10	7
	5:	20	10	15	10	3
Required:		4	5	8	9	

16.7 Computer Projects

1. Write a procedure to implement the column minima algorithm for finding an initial basic solution.

2. Using the results of the procedure treebuild, write a procedure to determine the value of θ and generate a new basic solution.

3. Develop a total package for solving the Simple Transportation Problem, with a suitable user interface. Your program should handle the case when the total available is not equal to the total required.

CHAPTER 17

The Assignment Problem

17.1 Introduction

The assignment problem is another deterministic OR problem which turns out to be a special case of the general LP problem. However, the way in which it can be solved is interesting and instructive.

Suppose that we have n men (without being sexist - "man" meaning "human being") and n jobs to be filled. Somehow or other we have measured the efficiency of each man at each job, and r_{ij} is the efficiency of the ith man at the jth job. How should men be assigned to jobs to maximise the total efficiency ? We can express the problem mathematically:

Let x_{ij} = 1 if man i does job j

$\qquad\quad$ = 0 otherwise.

Then maximise $Z = \sum_{i=1}^{n} \sum_{j=1}^{n} r_{ij} x_{ij}$
Subject to:

$\sum_{j=1}^{n} x_{ij} = 1$, $i = 1, \ldots n$ (Man i does one job)

$\sum_{i=1}^{n} x_{ij} = 1$, $j = 1, \ldots n$ (Job j is done by one man)

$\qquad x_{ij}$ = 0 or 1 for all i and j .

This is obviously closely similar to a Simple Transportation Problem, if we set $c_{ij} = -r_{ij}$ so that we turn it into a minimising problem. The only difference is in the last condition, $x_{ij} = 0$ or 1 instead of $x_{ij} \geq 0$. However, we know from the Simple Transportation algorithm that if we have integer values of $\{a_i\}$ and $\{b_j\}$, then we must obtain integer values of $\{x_{ij}\}$, since it involves only addition and subtraction, but no division. Because in this case x_{ij} cannot be greater than 1, the condition $x_{ij} \geq 0$ is equivalent to $x_{ij} = 0$ or 1.

Thus, the assignment problem is in fact identical to a Simple Transportation problem. A further interesting fact emerges from this - if we relax the conditions slightly, so that men may time-share jobs, then x_{ij} represents the fraction of time spent by man i on job j. As far as the mathematical formulation is concerned, this just means that the condition on x_{ij} is now $x_{ij} \geq 0$. But we have just shown that the optimal solution must have integer values, i.e. $x_{ij} = 0$ or 1. So the optimal solution is always "one man, one job".

An amusing variant is the marriage problem. We have n men (male humans this time) and n women, and r_{ij} measures the "happiness" when man i is with woman j. Let x_{ij} be the fraction of the time that man i spends with woman j. Then again the optimal solution which maximises the total happiness has $x_{ij} = 0$ or 1 - i.e. monogamy is best, proved mathematically. Note, however, that if you wish to minimise the total happiness the answer is also monogamy, although with different partners.

17.2 Basic Strategy for Solution

We could, of course, solve this problem by the Simple Transportation method described in the previous chapter, but there is in fact a simpler method tailored to the particular structure of this problem. A solution for the assignment problem is found when we have a set of n cells chosen from the $n \times n$ matrix of efficiencies $\{r_{ij}\}$ such that:

1. Every cell of the set is in a different row and column to
   every other cell of the set.
2. The sum of the efficiencies for the cells of the set is a
   maximum for all such sets.

   To find such a solution we adopt the following strategy:
transform the problem by stages into an equivalent problem with
the same solution set, but which is much easier to solve. Such
an equivalent problem is one with the following properties:

1. It is a *minimisation* problem, to find the smallest total Z
   value.
2. All values in the matrix are $\geq$ 0.
3. A set of n zero values can be found with no two values on
   the same row or column.

This set of zeroes is clearly going to be a minimum solution of
the transformed problem, and hence a maximum solution of the
original problem. The way in which we carry out the transformations depends on the following fact:

> *If we add or subtract a constant to or from any row*
> *or column of the matrix, then the optimal solution*
> *to the problem remains unchanged.*

Application of this principle systematically to the assignment
problem leads to the "Hungarian" solution method.

17.3 Hungarian Solution Method

   Before outlining the solution method, we need to make
some definitions.

   A set of elements of a square matrix is an *independent*
set if no two elements of the set lie on the same row or
column. We shall indicate a row or column of the matrix to be
covered by drawing a line through it.

   First, we shall use these definitions to state *Konig's*
theorem, although we shall not attempt to prove it.

   If A is a square matrix, then the maximum number of

independent zero elements of A is equal to the minimum number
of lines to be drawn to cover *all* the zero elements of A.

For example, if $A \;=\;$

$$\begin{pmatrix} 8 & 0^* & 4 & 2 & 0 \\ 2 & 0 & 5 & 0^* & 6 \\ 1 & 1 & 0^* & 1 & 4 \\ 3 & 2 & 0 & 2 & 1 \\ 0^* & 0 & 4 & 7 & 0 \end{pmatrix}$$

(The 0^* elements are
members of the
independent set of
zeroes).

> Maximum number of independent zeroes $=$ 4
> $=$ minimum number of covering lines.

To illustrate the solution method, we shall consider the
maximising assignment problem with the following matrix:

$$R \;=\; \begin{pmatrix} 52 & 73 & 85 & 75 & 48 \\ 32 & 51 & 70 & 60 & 62 \\ 46 & 62 & 38 & 60 & 54 \\ 71 & 54 & 42 & 50 & 55 \\ 60 & 48 & 35 & 50 & 36 \end{pmatrix}$$

Step 1

The first step is to convert the problem to an equivalent
minimisation one. To do this, we just multiply all the
elements of R by -1, since minimising $-Z$ is the same as
maximising Z. To make everything ≥ 0, we just add on the
largest r_{ij} value to every element in the matrix. In fact,
what is simplest is to subtract each element in turn from the
largest value.

Note that if the problem we have is a minimisation one to
start with, then we should omit step 1 and proceed straight to
step 2. Our example matrix after step 1 is:

$$\begin{pmatrix} 33 & 12 & 0 & 10 & 37 \\ 53 & 34 & 15 & 25 & 23 \\ 39 & 23 & 47 & 25 & 31 \\ 14 & 31 & 43 & 35 & 30 \\ 25 & 37 & 50 & 35 & 49 \end{pmatrix}$$
(Maximum value
was 85).

Step 2

Find the smallest value in each column and subtract it
from all the elements in that column.

$$\begin{pmatrix} 19 & 0 & 0 & 0 & 14 \\ 39 & 22 & 15 & 15 & 0 \\ 25 & 11 & 47 & 15 & 8 \\ 0 & 19 & 43 & 25 & 7 \\ 11 & 25 & 50 & 25 & 26 \end{pmatrix}$$

Step 3

Find the smallest value in each row and subtract it from all the elements in that row.

$$\begin{pmatrix} 19 & 0 & 0 & 0 & 14 \\ 39 & 22 & 15 & 15 & 0 \\ 17 & 3 & 39 & 7 & 0 \\ 0 & 19 & 43 & 25 & 7 \\ 0 & 14 & 39 & 14 & 15 \end{pmatrix}$$

We have now reached the stage where every row and column contains at least one zero element. The next task is to look for a maximum set of independent zeroes. If the number (k_M) of such zeroes is equal to n, the size of the matrix, then the positions of those zeroes will indicate an optimal solution. For our example, a maximum set is indicated below, with $k_M = 3$, together with 3 covering lines to cover all the zeroes.

$$\begin{pmatrix} 19 & 0^* & 0 & 0 & 14 \\ 39 & 22 & 15 & 15 & 0 \\ 17 & 3 & 39 & 7 & 0^* \\ 0^* & 19 & 43 & 25 & 7 \\ 0 & 14 & 39 & 14 & 15 \end{pmatrix}$$

In this case, it is clear that we do not yet have an optimal solution, and so more transformations are necessary.

Step 5

If $k_M < n$, we find the smallest uncovered element and call it h. Add h to every covered row or column, and subtract it from the whole matrix. Or, equivalently, add h to every twice-covered element, subtract it from each uncovered element, and leave once-covered elements unchanged. For our example, $h = 3$ and we get the following result:

$$\begin{pmatrix} \cancel{22} & \cancel{0} & \cancel{0*} & \cancel{0} & \cancel{17} \\ 39 & 19 & 12 & 12 & 0* \\ \cancel{17} & \cancel{0*} & \cancel{36} & \cancel{4} & \cancel{0} \\ 0* & 16 & 40 & 22 & 7 \\ 0 & 11 & 36 & 11 & 15 \end{pmatrix}$$

We find $k_M = 4$ for this matrix. This is not optimal, so we need to repeat step 5 until a matrix is obtained with the right number ($n = 5$) of independent zeroes. The smallest uncovered element $h = 11$, and we get:

$$\begin{pmatrix} 33 & 0 & 0* & 0 & 28 \\ 39 & 8 & 1 & 1 & 0* \\ 28 & 0* & 36 & 4 & 11 \\ 0* & 5 & 29 & 11 & 7 \\ 0 & 0 & 25 & 0* & 15 \end{pmatrix}$$

Now $k_M = 5$, and we have an optimal solution. Since we have transformed the original problem by adding and subtracting constants from rows and columns, the same solution will apply to the original problem. Taking the positions of the independent zeroes, and referring to the original matrix, we have the following solution:

$$\begin{aligned}
\text{Man 1 does Job 3} &: r_{13} = 85 \\
\text{Man 2 does Job 5} &: r_{25} = 62 \\
\text{Man 3 does Job 2} &: r_{32} = 62 \\
\text{Man 4 does Job 1} &: r_{41} = 71 \\
\text{Man 5 does Job 4} &: r_{54} = 50 \\
\text{Total efficiency} &\quad\;\, = 330.
\end{aligned}$$

One thing to note is that step 5 can only be repeated a finite number of times before an optimal solution is found. Each time we add a total of $k_M \times n \times h$ to the matrix, and subtract $n^2 \times h$. Since $k_M < n$, the sum of all the matrix elements must decrease each time. However, each element stays ≥ 0, so that ultimately a state in which all elements were zero would be reached. Before this, n independent zeroes would be found.

17.4 Finding Independent Zeroes

For assignment problems with n about 5, we can usually
spot the independent zeroes by eye. However, for larger
problems or for a computer solution of the problem, we need
to develop an algorithm to find independent zeroes and the
covering lines.

Let us assume that we have a square $n \times n$ matrix A with at
least one zero in every row and column, and that we have
already marked k independent zeroes in the matrix, with $k < n$.
The algorithm must terminate each time in one of two
alternatives:

Alternative 1: A new set of $k+1$ independent zeroes is marked.

Alternative 2: $k_M = k$; i.e. the maximum number of independent
zeroes is k.

We perform the algorithm repeatedly, starting at $k = 0$, until
either *Alternative 2* is reached or $k = n$.

Algorithm

Search each column of A for a 0*, until a column is found
with no 0* - call this the *pivotal* column, and note all the
rows which have zeroes in this column. These rows are searched
for 0*'s. Two things may transpire:

Either A row is found with no 0* in it. The 0 in this row and
the pivotal column can be marked as a 0*, and we have
reached *Alternative 1*.

Or   No such row is found. Start to make a list of row
numbers, starting with those rows that contain a 0 in the
pivotal column: $i_1, i_2, \ldots i_t$;
Add further terms to this list as follows: consider the
0* in row i_1 - add to the list all those row numbers
which are not already in the list and which contain a 0
in the same column as this 0*. Repeat for each row
number in the list, continuing after i_t. In this process

272

one of two things may transpire:

Either   A row (i_s) is reached which does not contain a 0*. Row
   i_s was "generated" by a previous row (i_r) because it
   contained a 0 in the same column as the 0* of row i_r.
   Transfer the * from row i_r to row i_s. Repeat this
   transfer process until a row is reached in the original
   sequence $i_1, \ldots i_t$. Mark the 0 in this row and the
   pivotal column as a 0*, and we have reached *Alternative 1.*

Or   We reach the end of the sequence and every row in it
   contains a 0*. We now move on to try all other possible
   pivotal columns, and if we get the same result from each
   one, we have reached *Alternative 2.*

This algorithm is more complicated to describe than it is
to carry out. To show this, let us examine an example, in
which we have already marked $k = 3$ independent zeroes:

$$\begin{pmatrix} 0* & 6 & 1 & 3 & 7 \\ 0 & 0* & 0 & 0 & 0 \\ 0 & 8 & 0* & 8 & 0 \\ 2 & 0 & 0 & 1 & 5 \\ 0 & 1 & 4 & 4 & 6 \end{pmatrix}$$

Pivotal column is 4. Form sequence:
                2 ;   (row with 0 in column 4).
Examine row 2 and find a 0* - add to the list other rows with
0's in the same column (2) :   2 ; 4 .
Row 4 has no 0* - shift the * from row 2 to row 4, and add an
extra * in row 2 and the pivotal column 4:

$$k = 4 \quad \begin{pmatrix} 0* & 6 & 1 & 3 & 7 \\ 0 & 0 & 0 & 0* & 0 \\ 0 & 8 & 0* & 8 & 0 \\ 2 & 0* & 0 & 1 & 5 \\ 0 & 1 & 4 & 4 & 6 \end{pmatrix}$$

Now column 5 is pivotal - the sequence is   2, 3 : 4 .
All these rows have 0*'s, and so we have reached *Alternative 2.*
Therefore, $k_M = 4$. We can now transform the matrix, using step
5 of the assignment algorithm.

$$k_M = 4$$
$$h = 1$$

$$\begin{pmatrix} 0^* & 6 & 1 & 3 & 7 \\ 0 & 0 & 0 & 0^* & 0 \\ 0 & 8 & 0^* & 8 & 0 \\ 2 & 0^* & 0 & 1 & 5 \\ 0 & 1 & 4 & 4 & 6 \end{pmatrix}$$

New matrix:

$$\begin{pmatrix} 0^* & 5 & 0 & 2 & 6 \\ 1 & 0 & 0 & 0^* & 0 \\ 1 & 8 & 0^* & 8 & 0 \\ 3 & 0^* & 0 & 1 & 5 \\ 0 & 0 & 3 & 3 & 5 \end{pmatrix}$$

Column 5 is pivotal - the sequence is 2, 3 ; 1, 4, 5. Row 5 has no 0*, so we shift the * from row 1 to row 5. Row 1 was generated in the list by row 3, so we shift the * from row 3 to row 1. Finally, we mark row 3, column 5 with a * :

$$\begin{pmatrix} 0 & 5 & 0^* & 2 & 6 \\ 1 & 0 & 0 & 0^* & 0 \\ 1 & 8 & 0 & 8 & 0^* \\ 3 & 0^* & 0 & 1 & 5 \\ 0^* & 0 & 3 & 3 & 5 \end{pmatrix}$$

$k = 5$, so we have an optimal solution.

17.5 Finding Covering Lines

To complete an automated system for the assignment problem, we need an algorithm for finding the k_M covering lines for carrying out step 5. Suppose we have reached *Alternative 2*, with $k_M < n$, and our last sequence of row numbers was $i_1, i_2, \ldots i_t$. Corresponding to each row i_r in this sequence is a column j_r, such that the 0* in row i_r is in column j_r. Then all the zeroes in the matrix can be covered by no more than the following $n-1$ lines:
1. All the rows $i_1, i_2, \ldots i_t$.
2. All the columns except $j_1, j_2, \ldots j_t$ and the pivotal column.

Example:

$$\begin{pmatrix} 0^* & 1 & 3 & 2 & 0 & 0 \\ 8 & 1 & 0^* & 1 & 0 & 3 \\ 5 & 0^* & 2 & 0 & 0 & 1 \\ 6 & 1 & 0 & 3 & 0^* & 2 \\ 5 & 9 & 1 & 1 & 0 & 4 \\ 0 & 4 & 3 & 0 & 1 & 0^* \end{pmatrix}$$

Column 4 is pivotal - the sequence is 3, 6 ; 1.
Alternative 2 reached. The corresponding columns are: 2, 6, 1.
Therefore, cover rows 1, 3 and 6 and columns 3 and 5 (see
above). Smallest uncovered element $h = 1$. Transformed matrix:

$$\begin{pmatrix} 0^* & 1 & 4 & 2 & 1 & 0 \\ 7 & 0 & 0^* & 0 & 0 & 2 \\ 5 & 0^* & 3 & 0 & 1 & 1 \\ 5 & 0 & 0 & 2 & 0^* & 1 \\ 4 & 8 & 1 & 0^* & 0 & 3 \\ 0 & 4 & 4 & 0 & 2 & 0^* \end{pmatrix}$$

This gives us an optimal solution.

17.6 Computer implementation

The Pascal procedure findzeroes makes use of the
following principal data structures:
<u>val</u> (2-dimensional integer array)
is the matrix of efficiency values.
<u>rowstar</u> (1-dimensional integer array)
stores the locations of the independent zeroes indexed by
columns: rowstar$[j]$ is the number of the row in which the
0^* in column j is to be found (= 0 if there is no 0^*).
<u>colstar</u> (1-dimensional integer array)
stores the locations of the independent zeroes indexed by
rows: colstar$[i]$ is the number of the column in which the 0^*
in row i is to be found (= 0 if there is no 0^*).
(Note: this way of storing the information, by both rows and
columns, leads to a certain amount of redundancy, but makes
the program simpler and more efficient).

list (1-dimensional integer array)

      stores the list of row numbers for the algorithm.

listed (1-dimensional boolean array)

      tells whether or not a row is in the list.

```
procedure findzeroes(var val : matrix; var rowstar,colstar,
                     list : intarr; var nlist,km,n : posint);
{ Procedure to find independent zeroes of a square matrix. }
var i,j,k,pivot : posint;
    staradded : boolean;
    pred : intarr;
begin
  for i := 1 to n do
  begin
    rowstar[ i ] := 0; colstar[ i ] := 0;
  end;
  repeat
    staradded := false; pivot := 0;
    repeat
      pivot := pivot + 1;
      if rowstar[ pivot ] = 0
      then begin
        nlist := 0; for i := 1 to n do listed[ i ] := false;
        for i := 1 to n do
          if val[ i,pivot ] = 0
          then begin
            nlist := nlist + 1; list[ nlist ] := i;
            pred[ nlist ] := 0; listed[ i ] := true;
          end;
        k := 0;
        repeat
          k := k + 1;
          if colstar[ list[ k ] ] > 0
          then begin
            j := colstar[ list[ k ] ];
            for i := 1 to n do
              if (val[ i,j ] = 0) and not listed[ i ]
              then begin
                nlist := nlist + 1; list[ nlist ] := i;
                pred[ nlist ] := k; listed[ i ] := true;
              end
          end
          else begin
            while pred[ k ] > 0 do
            begin
              j := colstar[ list[ pred[ k ] ] ];
              colstar[ list[ k ] ] := j;
              rowstar[ j ] := list[ k ]; k := pred[ k ];
            end;
            rowstar[ pivot ] := list[ k ];
            colstar[ list[ k ] ] := pivot;
            km := km + 1; staradded := true;
          end;
        until (k = nlist) or staradded;
```

```
        end;
    until (pivot = n) or staradded;
    writeln; writeln('km = ',km:4);
    for i := 1 to n do
        writeln(i:4,rowstar[ i ],colstar[ i ]);
    until (not staradded) or (km = n);
end;
```

17.8 Exercises

1. A modern torture chamber uses Operations Research to
    optimise its results. The Chief Torturer has 5 prisoners
    to work on, and he has drawn up a table (in arbitrary
    units) expressing how much agony he believes each of his
    5 instruments of torture will cause each victim:

Victims	Rack	Thumbscrews	Pincers	Boiling Oil	Chinese Water Torture
A	5	9	4	3	6
B	7	7	1	8	5
C	3	4	5	2	5
D	5	6	2	3	7
E	2	8	1	2	5

    How should he allocate his tortures to cause maximum
    agony ? His Assistant Torturer is secretly sympathetic to
    the prisoners, and re-arranges the schedule to minimise
    the victims' agony. How should he do this ?

2. An oil company wishes to move its 5 oil rigs from their
    present locations to drill 5 new wells. The distances
    from each present site to each new location are as
    follows, in miles:

		A	B	C	D	E
Old	1	19	20	25	28	7
Sites	2	53	49	50	46	51
	3	39	37	41	38	38
	4	53	55	52	54	53
	5	50	47	52	48	49

New Sites

How should they move their rigs to minimise the total distance travelled ?

3. 6 men and 6 women are standing on a square dance-floor.
Their coordinates (measured in metres north and east from one corner of the dance-floor) are:

Arthur	(3,8)	Anne	(2,9)
Bill	(3,7)	Betty	(9,9)
Charlie	(6,8)	Clara	(9,8)
Dave	(7,4)	Daisy	(9,6)
Edgar	(5,2)	Ethel	(6,5)
Fred	(8,2)	Flo	(7,1)

When the music starts, the men each select a woman and move towards her. The embarassment caused is proportional to the square of the distance from man to woman. Which woman should they each pick to minimise the total embarassment, so that each man gets one and only one woman ?

17.9 Computer Projects

1. Write a procedure which, following on from findzeroes, determines the covering rows and columns for the matrix.

2. Write a program using these procedures to solve the general assignment problem, for either maximising or minimising the objective function.

CHAPTER 18

Games Theory

18.1 Two-Person Zero-Sum Games

Games theory is concerned with the problem of determining
a strategy for the player of a game which will bring the highest returns in the long run, assuming that both players use
their best strategies. By "players" we may mean individuals,
or groups of individuals, or corporations, and by "game" we
mean any situation where interests conflict. In this chapter
we shall confine ourselves to two-person, zero-sum games. By
this we mean:

1. Two-person - there are only two players, or interest
   groups, involved in the game.
2. Zero-sum - what one player wins, the other loses, and vice
   versa. The sum of the winnings of both players is always
   zero.

For any such game, between say players A and B, we shall
specify a *pay-off matrix*, which represents the amount that A
wins from B for each combination of moves by A and B. Assume
that each player has only a finite number of possible moves
available, and that the game will be repeated indefinitely with
the same pay-off matrix.

Example: Pay-off matrix.

B

	1	2	3	4	
1	3	-2	1	-3	*-3*
2	-2	-2	-1	0	*-2*
3	1	-1	0	0	*(-1)*
4	-1	-2	-3	2	*-3*

(A is to the left of the rows; handwritten below columns: *3* *(-1)* *1* *2*)

A is pessimistic, and looks to see what the worst result is for each of his 4 moves, by looking at the minimum value in each row. These are -3, -2, -1 and -3. Therefore, if he plays move 3 he is guaranteed to do no worse than -1, whereas with any other move he could do worse. In other words, the maximum of the row minima is -1.

In the same way, B looks to see what the worst value is for each of his 4 moves - in other words, he looks at the maximum value in each column, which come to 3, -1, 1 and 2. So if B plays move 2 every time, he is guaranteed to do no worse than -1, which is the minimum of the column maxima.

But now both A and B can guarantee the same result by playing the same move each time, because the game has what is known as a *saddle-point* - i.e. the maximum of the row minima is equal to the minimum of the column maxima. If either player departs from his best move, then the other player will do better. So both players have what are known as *pure strategies*, consisting of just a single move. We write these strategies as: A(0,0,1,0) and B(0,1,0,0). The *value* of the game is the amount that A expects to win from B each time they play, and in this case it is -1.

Games with saddle-points are very boring and easy to analyse. In general, games will not have saddle-points and the solutions will not be pure strategies.

18.2 Mixed Strategies

To study the situation where no saddle-point exists, we shall consider initially just 2×2 games, in which each player has just two moves. For example:

```
                    B
            1       2       Row minima

        1 |  3  | -4  |       -4
   A      |-----|-----|
        2 | -2  |  3  |       -2

Column max.  3       3
```

The maximum of the row minima is -2 and the minimum of the column maxima is 3, so there is no saddle-point. The value of the game is clearly between -2 and 3, but equally clearly no single pure strategy will achieve it. If either player always makes the same moves, or makes his move in any kind of discernible pattern, then the other player will eventually spot this and adjust his move to maximise his own winnings.

The only sensible thing to do is to choose moves at random, so that the next move is unpredictable by the other player. Suppose A plays move 1 with probability p and move 2 with probability $1-p$. In general his best strategy will be to ensure that his expected winnings against each of B's two moves are the same.

Against B's first move his expected winnings are $3p - 2(1-p)$.
Against B's second move they are $-4p + 3(1-p)$.
Therefore, the best value of p is the one for which

$$3p - 2(1-p) = -4p + 3(1-p)$$
$$\text{or} \quad p = 5/12 .$$

A simple way of computing these probabilities for both A and B, but which only works for 2×2 games, is called the *method of oddments*. This works as follows:

1. Calculate the absolute difference between the pay-off values in each row and write the result against the *other row*.
2. Use these values as the odds in favour of each move.
3. Repeat for the columns to give the odds for B's moves.

For example:

		B		Differences
		1	2	
A	1	3	-4	5
	2	-2	3	7
Differences		7	5	

So the odds for A's move 1 are 5:7, or a probability of 5/12. The final solution gives the probabilities for each move of each player: A(5/12, 7/12) , B(7/12, 5/12) . The value of the game is the expected winnings by A against either of B's moves = 3×5/12 - 2×7/12 = -4×5/12 + 3×7/12 = 1/12 . Therefore this game is biased in favour of A by this amount.

This kind of solution is called a *mixed strategy* because each player no longer plays the same move each time but a mixture of moves, determined by a probability distribution.

18.3 Dominance

We have seen how to find the mixed strategy for a 2×2 game by the method of oddments, but this method does not work for larger games. If we have a game where each player has more than 2 moves, we may be able to reduce it to a smaller, more easily solved, problem. For example, consider the game with the following pay-off matrix:

		B				
		1	2	3	4	5
	1	-1	-2	4	2	0
A	2	2	3	-5	-1	-3
	3	-1	-3	4	1	-1

The maximum of the row minima is -2, and the minimum of the column maxima is 0, so the game has no saddle-point. Consider A's move 3 by comparison with his move 1. Whatever move B makes, A never gets a better result by playing move 3 rather than move 1. A would never play move 3, therefore, and we can forget about it. We say that A's move 1 *dominates* move 3. Deleting move 3, we get a smaller game:

		B				
		1	2	3	4	5
A	1	-1	-2	4	2	0
	2	2	3	-5	-1	-3

Considering now B's moves, and comparing them with each other in turns, we see that move 4 is always worse for B than move 5 (the pay-off to A is larger for both of A's moves). Therefore, move 5 dominates move 4, and we can delete the latter from the game.

		B			
		1	2	3	5
A	1	-1	-2	4	0
	2	2	3	-5	-3

We are left with no more dominated moves, and thus the size of the game can be reduced no further. However, by considering dominance we have reduced the size of the game from 3×5 to 2×4. We can still not solve this by the method of oddments, and so must turn to other techniques.

18.4 2xN Games and Linear Programming

For the general 2×N game, where N is bigger than 2, we can formulate an associated LP problem. We wish to determine two unknown quantities: p, the probability with which A plays his first move, and V, the value of the game to A. Now A wants to maximise the value of the game, subject to the condition that whatever move B makes the expected winnings are always

greater than or equal to V. For example, considering the game
we were left with at the end of the last section, if B plays
move 1, A expects to win $-1p + 2(1-p) \geq V$.
Similarly, for each one of B's moves we can write down the
associated condition:

$$-2p + 3(1-p) \geq V \quad \text{(Move B2)},$$
$$4p - 5(1-p) \geq V \quad \text{(Move B3)},$$
$$0p - 3(1-p) \geq V \quad \text{(Move B5)}.$$

It must also be true that $0 \leq p \leq 1$, since p is a
probability.

So, looking at everything from A's point of view, we have
the following LP problem:

Maximise V

$$\begin{aligned}
\text{Subject to:} \quad 3p + V &\leq 2 \\
5p + V &\leq 3 \\
-9p + V &\leq -5 \\
-3p + V &\leq -3 \\
p &\leq 1 \\
p &\geq 0
\end{aligned}$$

This LP problem has just two variables, p and V, and can
therefore be solved graphically.

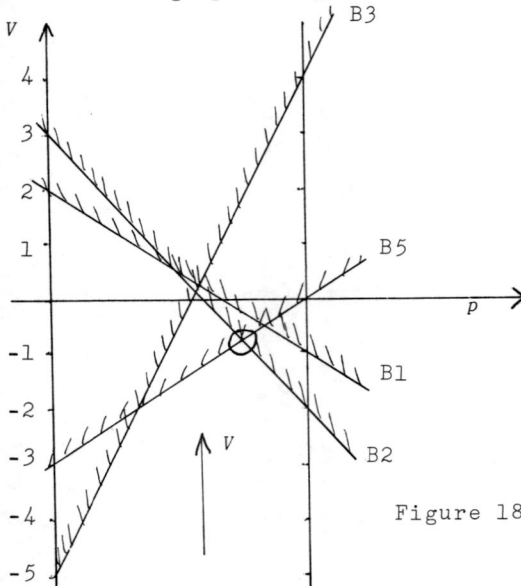

Figure 18.1

284

In fact, drawing the constraints for this problem is very simple - mark two parallel vertical lines with the corresponding pay-offs for A's two moves for each move of B's, and join each such pair to form a constraint line. For example, the constraint corresponding to B's move 1 joins the point 2 on the $p = 0$ line to the point -1 on the $p = 1$ line, and so forth.

In this case, it is clear that the maximum value of V is at the intersection of B2 and B5. To find the actual solution, the simplest thing to do is to solve the resulting 2×2 game by the method of oddments:

Solution: A(3/4, 1/4, 0), B(0, 3/8, 0, 0, 5/8). Value = -3/4.

Any 2×N game can be solved graphically in a similar way. To solve an N×2 game, in which B has only 2 moves and A has several, two approaches may be adopted. Either A and B can be interchanged, and by multiplying all the pay-off values by -1 a 2×N game is produced which can be solved in the same way as above. Or we may write down a similar LP problem, looking at the game from B's point of view, remembering that B wishes to minimise the value and to ensure that the expected winnings for each of A's moves is less than or equal to V. An identical result is achieved in either case.

The overall strategy for tackling two-person, zero-sum games can now be outlined in the following stages:
1. Look for a saddle-point. If one exists, there is a pure strategy and the game is solved.
2. Otherwise, remove all dominated rows and columns until no such moves remain in the pay-off matrix.
3. If the result is a 2×2 game, solve by the method of odd-ments.
4. If it is a 2×N or N×2 game, set up a 2-variable LP problem and solve graphically.
5. If it is bigger than this, it can still be formulated as an LP problem, but cannot be solved graphically.

18.5 Exercises

1. The sun is setting as the farmer and the fox face each other across the river. The farmer has 5 geese with him and a small boat. He knows that during the night he can cross the river with any number of geese, or he can stay on the side where he is with all the geese. The fox can either stay on the side where he is, or he can cross the river during the night. In the morning, the fox will attack any geese who are on the same side of the river. If they are alone, he will kill them all. If the farmer is with them, the expected number killed is half the number present. What are the best strategies for the fox and the farmer ?

2. Solve the following games by finding the optimal strategies for A and B and the value of the game:

a)

	B 1	2	3
1	3	0	2
A 2	-4	-1	3
3	2	-2	-1

b)

	B 1	2	3
1	3	0	2
A 2	4	5	1
3	2	3	-1

c)

	B 1	2	3	4	5
1	2	5	-1	0	8
2	0	-2	7	3	-4
A 3	1	5	-2	-1	6
4	0	-3	6	1	-5

3. Fred and Bert play a simple game. Fred holds out either one or two fingers, and simultaneously Bert holds out

either one or two fingers and shouts out a number. If the number equals the total number of fingers held out, Fred pays Bert as many pounds. Otherwise, Bert pays Fred a sum in pounds equal to twice the number of fingers Bert held out. Find the optimal strategies and the value of the game to Bert.

4. A hunter with a rifle is facing a lion which is pacing towards him. If the hunter decides to fire at a range of 15 feet, there is a 25% chance that he will kill the lion. If he misses, he will have time to reload by the time the lion has stalked to 4 feet away (if it has not leapt first) and then he will have a 75% chance of killing it. If he fires from 9 feet, there is an even chance of killing it, but no time to reload.

   If the lion decides to leap on the man from 12 feet, there is an even chance of the hunter escaping. If the lion leaps from 6 feet, the hunter has a 25% chance of escaping, and if it leaps from 3 feet there is no escape. Determine the optimal strategies for man and lion, and the probability of the hunter escaping.
   (Use as pay-off the probability of the hunter escaping for each combination of moves).

18.6 Computer Projects

1. Write a program to detect a saddle-point of a game, and output the pure strategy, if one exists.

2. Write a program to solve a 2×2 game by the method of oddments.

3. Write a program to detect dominated rows and columns.

4. Write a program to solve a 2×N or N×2 game. Combine this with the others to produce a general package.

Suggested Further Reading for Section IV

"Principles of Operations Research", by Harvey M. Wagner, Prentice-Hall, 1975.

"Operational Research", by S.S. Cohen, Edward Arnold, 1985.

"Linear Programming - A Computational Approach", by K-K. Lau, Chartwell-Bratt, 1984.

"Linear Programming", by G. Hadley, Addison-Wesley, 1962.

"An Introduction to Linear Programming", by G.R. Walsh, Holt, Rinehart & Winston, 1971.

"An Introduction to Linear Programming and Matrix Game Theory", by M.J. Fryer, Edward Arnold, 1978.

"The Compleat Strategyst", by J.D. Williams, McGraw-Hill, 1954.

APPENDIX A

Statistical Tables

Table 1 - Standard Normal Table

The values in the table are the probabilities of the random variable Z lying between 0 and z (or -z and 0).

z	.00	.01	.02	.03	.04	.05	.06	.07	.08	.09
0	.0000	.0040	.0080	.0120	.0160	.0199	.0239	.0279	.0319	.0359
.1	.0398	.0438	.0478	.0517	.0557	.0596	.0636	.0675	.0714	.0753
.2	.0793	.0832	.0871	.0909	.0948	.0987	.1026	.1064	.1103	.1141
.3	.1179	.1217	.1255	.1293	.1331	.1368	.1406	.1443	.1480	.1517
.4	.1555	.1591	.1628	.1664	.1700	.1736	.1772	.1808	.1844	.1879
.5	.1915	.1950	.1985	.2019	.2054	.2088	.2123	.2157	.2190	.2224
.6	.2257	.2291	.2324	.2357	.2389	.2422	.2454	.2486	.2517	.2549
.7	.2580	.2611	.2642	.2673	.2703	.2734	.2764	.2794	.2822	.2852
.8	.2881	.2910	.2939	.2967	.2995	.3023	.3051	.3078	.3106	.3133
.9	.3159	.3186	.3212	.3238	.3264	.3289	.3315	.3340	.3365	.3389
1.0	.3413	.3438	.3461	.3485	.3508	.3531	.3554	.3577	.3599	.3621
1.1	.3643	.3665	.3686	.3708	.3729	.3749	.3770	.3790	.3810	.3830
1.2	.3849	.3869	.3888	.3907	.3925	.3944	.3962	.3980	.3997	.4015
1.3	.4032	.4049	.4066	.4082	.4099	.4115	.4131	.4147	.4162	.4177
1.4	.4192	.4207	.4222	.4236	.4251	.4265	.4279	.4292	.4306	.4319
1.5	.4332	.4345	.4357	.4370	.4382	.4394	.4406	.4418	.4429	.4441
1.6	.4452	.4463	.4474	.4484	.4495	.4505	.4515	.4525	.4535	.4545
1.7	.4554	.4564	.4573	.4582	.4591	.4599	.4608	.4616	.4625	.4633
1.8	.4641	.4649	.4656	.4664	.4671	.4678	.4686	.4693	.4699	.4706
1.9	.4713	.4719	.4726	.4732	.4738	.4744	.4750	.4756	.4761	.4767
2.0	.4772	.4778	.4783	.4788	.4793	.4798	.4803	.4808	.4812	.4817
2.1	.4821	.4826	.4830	.4834	.4838	.4842	.4846	.4850	.4854	.4857
2.2	.4861	.4865	.4868	.4871	.4875	.4878	.4881	.4884	.4887	.4890
2.3	.4893	.4896	.4898	.4901	.4904	.4906	.4909	.4911	.4913	.4916
2.4	.4918	.4920	.4922	.4925	.4927	.4929	.4931	.4932	.4934	.4936
2.5	.4938	.4940	.4941	.4943	.4946	.4947	.4948	.4949	.4951	.4952
2.6	.4953	.4955	.4956	.4957	.4959	.4960	.4961	.4962	.4963	.4964
2.7	.4965	.4966	.4967	.4968	.4969	.4970	.4971	.4972	.4973	.4974
2.8	.4974	.4975	.4976	.4977	.4977	.4978	.4979	.4979	.4980	.4981
2.9	.4981	.4982	.4982	.4983	.4984	.4984	.4985	.4985	.4986	.4986

Table 2 - Critical Points of the t Distribution

Degrees of Freedom	Error Probabilities							
	50%	25%	10%	5%	2.5%	1%	0.5%	0.1%
1	1.00	2.41	6.31	12.7	25.5	63.7	127	637
2	0.82	1.60	2.92	4.30	6.21	9.92	14.1	31.6
3	0.76	1.42	2.35	3.18	4.18	5.84	7.45	12.9
4	0.74	1.34	2.13	2.78	3.50	4.60	5.60	8.61
5	0.73	1.30	2.01	2.57	3.16	4.03	4.77	6.86
6	0.72	1.27	1.94	2.45	2.97	3.71	4.32	5.96
7	0.71	1.25	1.89	2.36	2.84	3.50	4.03	5.40
8	0.71	1.24	1.86	2.31	2.75	3.36	3.83	5.04
9	0.70	1.23	1.83	2.26	2.68	3.25	3.69	4.78
10	0.70	1.22	1.81	2.23	2.63	3.17	3.58	4.59
11	0.70	1.21	1.80	2.20	2.59	3.11	3.50	4.44
12	0.69	1.21	1.78	2.13	2.56	3.05	3.43	4.32
13	0.69	1.20	1.77	2.16	2.53	3.01	3.37	4.22
14	0.69	1.20	1.76	2.14	2.41	2.98	3.33	4.14
15	0.69	1.20	1.75	2.13	2.49	2.95	3.29	4.07
16	0.69	1.19	1.75	2.12	2.47	2.92	3.25	4.01
17	0.69	1.19	1.74	2.11	2.46	2.90	3.22	3.96
18	0.69	1.19	1.73	2.10	2.44	2.88	3.20	3.92
19	0.69	1.19	1.73	2.09	2.43	2.86	3.17	3.88
20	0.69	1.18	1.72	2.09	2.42	2.85	3.15	3.85
21	0.69	1.18	1.72	2.08	2.41	2.83	3.14	3.82
22	0.69	1.18	1.72	2.07	2.41	2.82	3.12	3.79
23	0.68	1.18	1.71	2.07	2.40	2.81	3.10	3.77
24	0.68	1.18	1.71	2.06	2.39	2.80	3.09	3.74
25	0.68	1.18	1.71	2.06	2.38	2.79	3.08	3.72
26	0.68	1.18	1.71	2.06	2.38	2.78	3.07	3.71
27	0.68	1.18	1.70	2.05	2.37	2.77	3.06	3.69
28	0.68	1.17	1.70	2.05	2.37	2.76	3.05	3.67
29	0.68	1.17	1.70	2.05	2.36	2.76	3.04	3.66
30	0.68	1.17	1.70	2.04	2.36	2.75	3.03	3.65
40	0.68	1.17	1.68	2.02	2.33	2.70	2.97	3.55
120	0.68	1.16	1.67	2.00	2.30	2.66	2.91	3.46
∞	0.67	1.15	1.64	1.96	2.24	2.58	2.81	3.29

The values in the table are for a two-sided test - in other words, they are the values of T such that the probability of being $> T$ or $< -T$ is equal to the specified error probability for a given number of degrees of freedom.

The last row corresponds to the Standard Normal random variable.

Table 3 - Critical Points of the χ^2 Distribution

Degrees of Freedom	Error Probabilities						
	99%	95%	90%	50%	10%	5%	1%
1	0.0001	0.004	0.016	0.455	2.705	3.841	6.635
2	0.020	0.103	0.211	1.386	4.605	5.991	9.210
3	0.115	0.352	0.584	2.366	6.251	7.815	11.34
4	0.297	0.710	1.064	3.357	7.779	9.488	13.28
5	0.554	1.145	1.610	4.351	9.236	11.07	15.09
6	0.872	1.635	2.204	5.348	10.64	12.59	16.81
7	1.239	2.167	2.833	6.346	12.02	14.07	18.48
8	1.646	2.733	3.490	7.344	13.36	15.51	20.09
9	2.088	3.325	4.168	8.343	14.68	16.92	21.67
10	2.558	3.940	4.865	9.342	15.99	18.31	23.21
11	3.053	4.575	5.578	10.34	17.28	19.68	24.73
12	3.571	5.226	6.304	11.34	18.55	21.03	26.22
13	4.107	5.892	7.042	12.34	19.81	22.36	27.69
14	4.660	6.571	7.790	13.34	21.06	23.68	29.14
15	5.229	7.261	8.547	14.34	22.31	25.00	30.58
16	5.812	7.962	9.312	15.34	23.54	26.30	32.00
17	6.408	8.672	10.09	16.34	24.77	27.59	33.41
18	7.015	9.390	10.86	17.34	25.99	28.87	34.81
19	7.633	10.12	11.65	18.34	27.20	30.14	36.19
20	8.360	10.85	12.44	19.34	28.41	31.41	37.57
22	9.542	12.34	14.04	21.34	30.81	33.92	40.29
24	10.86	13.85	15.66	23.34	33.20	36.42	42.98
26	12.20	15.38	17.29	25.34	35.56	38.89	45.64
28	13.56	16.93	18.94	27.34	37.92	41.34	48.28
30	14.95	18.49	20.60	29.34	40.26	43.77	50.89
40	22.16	26.51	29.05	39.34	51.81	55.76	63.69
50	29.71	34.76	37.69	49.33	63.17	67.50	76.15
60	37.48	43.19	46.46	59.33	74.40	79.08	88.38
70	45.44	51.74	55.33	69.33	85.53	90.53	100.4
80	53.54	60.39	64.28	79.33	96.58	101.9	112.3
90	61.75	69.13	73.29	89.33	107.6	113.1	124.1
100	70.06	77.93	82.36	99.33	118.5	124.3	135.8

These are the values of χ^2 such that the probability of being $> \chi^2$ is equal to the specified error probability for a given number of degrees of freedom.

Table 4 - Critical Points of the f Distribution

ν_2	\multicolumn{9}{c}{ν_1}
	1	2	3	4	5	7	10	24	∞
1	161	200	216	225	230	237	242	249	254
	4052	5000	5403	5625	5764	5928	6056	6235	6366
2	18.5	19.0	19.2	19.3	19.4	19.4	19.4	19.5	19.5
	98.5	99.0	99.2	99.2	99.3	99.4	99.4	99.5	99.5
3	10.1	9.55	9.28	9.12	9.01	8.89	8.79	8.64	8.53
	34.1	30.8	29.5	28.7	28.2	27.7	27.2	26.6	26.1
4	7.71	6.94	6.59	6.39	6.26	6.09	5.96	5.77	5.63
	21.2	18.0	16.7	16.0	15.5	15.0	14.5	13.9	13.5
5	6.61	5.79	5.41	5.19	5.05	4.88	4.74	4.53	4.36
	16.26	13.27	12.06	11.39	10.97	10.46	10.05	9.47	9.02
6	5.99	5.14	4.75	4.53	4.39	4.21	4.06	3.84	3.67
	13.74	10.92	9.78	9.15	8.75	8.26	7.87	7.31	6.88
7	5.59	4.74	4.35	4.12	3.97	3.79	3.64	3.41	3.23
	12.25	9.55	8.45	7.85	7.46	6.99	6.62	6.07	5.65
8	5.32	4.46	4.07	3.84	3.69	3.50	3.35	3.12	2.93
	11.26	8.65	7.59	7.01	6.63	6.18	5.81	5.28	4.86
9	5.12	4.26	3.86	3.63	3.48	3.29	3.14	2.90	2.71
	10.56	8.02	6.99	6.42	6.06	5.61	5.26	4.73	4.31
10	4.96	4.10	3.71	3.48	3.33	3.14	2.98	2.74	2.54
	10.04	7.56	6.55	5.99	5.64	5.20	4.85	4.33	3.91
12	4.75	3.89	3.49	3.26	3.11	2.91	2.75	2.51	2.30
	9.33	6.93	5.95	5.41	5.06	4.64	4.30	3.78	3.36
14	4.60	3.74	3.34	3.11	2.96	2.76	2.60	2.35	2.13
	8.86	6.51	5.56	5.04	4.70	4.28	3.94	3.43	3.00
16	4.49	3.63	3.24	3.01	2.85	2.66	2.49	2.24	2.01
	8.53	6.23	5.29	4.77	4.44	4.03	3.69	3.18	2.75
18	4.41	3.55	3.16	2.93	2.77	2.58	2.41	2.15	1.92
	8.29	6.01	5.09	4.58	4.25	3.84	3.51	3.00	2.57
20	4.35	3.49	3.10	2.87	2.71	2.51	2.35	2.08	1.84
	8.10	5.85	4.94	4.43	4.10	3.70	3.37	2.86	2.49
28	4.20	3.34	2.95	2.71	2.56	2.36	2.19	1.91	1.65
	7.64	5.45	4.57	4.07	3.75	3.36	3.03	2.52	2.06
40	4.08	3.23	2.84	2.61	2.45	2.25	2.08	1.79	1.51
	7.31	5.18	4.31	3.83	3.51	3.12	2.80	2.29	1.80
60	4.00	3.15	2.76	2.53	2.37	2.17	1.99	1.70	1.39
	7.08	4.98	4.13	3.65	3.34	2.95	2.63	2.12	1.60
∞	3.84	3.00	2.60	2.37	2.21	2.01	1.83	1.52	1.00
	6.63	4.61	3.78	3.32	3.02	2.64	2.32	1.79	1.00

ν_1 is the number of degrees of freedom for the top variance, and ν_2 is the number for the bottom. The upper critical value is for a 5% significance level, and the lower for 1%.

APPENDIX B

Answers to Numerical Exercises

Chapter 1

1. a) {1,2,3,4,5,6,7,8,10}
   b) {2}
   c) {9}
   d) {1,3,4,5,6,7,8,9,10}
   e) {2,4}

Chapter 2

1. a) $8x^3$ b) $-1/x^2$ c) $1/(2\sqrt{x})$

2. a) $(9+4x)/(2\sqrt{(9x+2x^2)})$
   b) $-3/(3x+1)^2$
   c) $(2x^4-3x^2)/(2x^2-1)^2$
   d) $-1/x$

3. $56\frac{1}{4}$ miles. 15 hours.

6. £3,750 in project A, rest in project B.

7. a) $y = x^2 + \dfrac{4x^3}{3} + c$

   b) $y = -\dfrac{1}{2x} + c$

   c) $y = \ln(1+x) + c$
   d) $y = -\frac{1}{2}exp(-2x) + c$

8. a) 8 b) 0 c) 64/3 d) e^2-e

9. 500/3

10. 3.2 secs.

11. Trapezoidal: 0.7703
    Simpson: 0.7726

Chapter 3

1. a) × b) $(\begin{matrix} 5 & 0 & 1 & 5 \end{matrix})$
   c) $\begin{pmatrix} -1 & -3 & 2 & -7 \\ 4 & -3 & -2 \\ 15 & 0 & 6 & 15 \end{pmatrix}$
   d) $\begin{pmatrix} 5 \\ 14 \end{pmatrix}$ e) ×

2. a) 32 b) $\begin{pmatrix} 11 & 12 \\ 12 & 33 \end{pmatrix}$

   c) $\begin{pmatrix} 13 & 11 & -7 & 6 \\ 11 & 17 & -9 & 12 \\ -7 & -9 & 5 & -6 \\ 6 & 12 & -6 & 9 \end{pmatrix}$

3. a) $\begin{pmatrix} \frac{3}{4} & \frac{1}{4} \\ \frac{1}{2} & \frac{1}{2} \end{pmatrix}$

   b) $\begin{pmatrix} \frac{1}{2} & 0 & -\frac{1}{2} \\ 0 & 1 & 0 \\ \frac{1}{2} & 0 & \frac{1}{2} \end{pmatrix}$

   c) Singular.

Chapter 4

1. a) 1/4 b) 2/13 c) 1/13
   d) 17/52 e) 1/2 f) 4/13

2. a) 17/36 b) 19/36 c) 7/12
   d) 23/39 e) Yes

3. a) 0.34 b) 0.27 c) 0.66
   d) 0.9296

4. 2,598,960. 1/54145.
   33/16660.

5. a) 1/45 b) 1/15 c) 2/15
   d) 1/15 e) 8/45 f) 2/3

6. a) 0.00553   b) 0.2903
   c) 0.7102

7. 0.6513

8. a) 0.133   b) 4

9. a) 5   b) 24

10. a)

x	$P(x)$
0	1/70
1	16/70
2	36/70
3	16/70
4	1/70

   b) 2   c) 0.756   d) 0.36

11. 0.1429

12. 0.4745

13. $F(x) = 1 - 1/x$.   1/3.
    Mean = ∞

14. $b = 3/32$.   Mean = 2.

15. 0.6915

16. a) 5.5%   b) 67.3%

17.   71.45, 61.75, 52.47,
      46.58, 42.19

18. a) 0.324   b) 0.463

Chapter 5

1. $\hat{\theta} = 2x$

2. $\bar{x} = 16$, $s^2 = 16.8$

3. (59.27, 64.73)
   (59.94, 64.06)
   (2.38, 5.67)

4. (0.0125, 0.02085).   90,660.

5. (110.3, 149.7). ~~1025.~~
   70,256

Chapter 6

1. 4 or 5

2. 5 or 6

3. Accept H_1 if $x < 2.565$

4. $Z = 2.767$.   Accept H_1.

5. Acceptance region
   (32.96, 37.04). Accept H_0.

6. $Z = 0.6$: Accept H_0
   $Z = 3.6$: Accept H_1

Chapter 7

1. $t = 2.494$: Accept H_1

2. $t = -4.585$: Accept H_1

3. $t = 3.475$: Accept H_1

4. $t = 6.912$: Accept H_1

5. $t = 5.4006$: Accept H_1

6. $t = 6.202$: Accept H_1

Chapter 8

1. $\chi^2 = 2.41$: Accept H_0

2. $\chi^2 = 1.778$: Accept H_0

3. a) $\chi^2 = 20.35$: Accept H_1
   b) $\chi^2 = 3.87$: Accept H_0

4. $\chi^2 \simeq 2.11$: Accept H_0

5. $\chi^2 = 4.24$: Accept H_0

Chapter 9

1. Oil: $Y = 1909.0 + 0.2877X$
   Water: $Y = -2194.7 + 0.7244X$
   Oil/water contact at 9397.2

2. a) $Y = 1.421 + 0.1828X^2$
   b) $Y = 1.194X^{1.1648}$
   c) $Y = 1.935(1.257)^X$

3. $Y = 2.8368 + 0.6864X$
   $X = -0.7083 + 1.0833Y$

4. $a = 2.941$, $b = 1968.0$.
   $t = 6.49$: Accept H_1

Chapter 10

1. a) $f = 5.195$: Not sig.
   b) $f = 258$: Significant.

2. $f = 12.08$: Significant

3. $f = 29.166$: Significant

4. $f = 11.84$: Significant

5. f values: 7.67, 3.07.
   Both significant.

Chapter 11

1. 0.3872, 0.1128

2. $^{2n}C_{n+a}\,p^{n+a}q^{n-a}$

   $\dfrac{a}{n}\,^{2n}C_{n+a}(\tfrac{1}{2})^{2n}$

3. $^{2n}C_{n+b-a} p^{n+b-a} q^{n-b+a}$

5. 0.0046.

Chapter 12

1. $\alpha = ((q/p)^a - (q/p)^{-b})^{-1}$
   $\beta = -\alpha(q/p)^{-b}$
   63/1023

2. 6 days 22 hours 40 mins.

3. $\tau(i) = -\dfrac{1}{2p} i^2 + \dfrac{a-b}{2p} + \dfrac{ab}{2p}$
   1000 minutes.

4. About 20%

5. $\tau(i) = (a-i)/(p+2q)$

Chapter 13

3. a) (2/9, 1/2, 5/18)
   b) (64/175, 45/175, 66/175)

4. (0.32, 0.48, 0.2)
   (8/27, 1/3, 10/27)

5. (1/16, 1/4, 3/8, 1/4, 1/16)

6. 8/5, 4/5, 4/5.
   P(Error) = 2/5.

Chapter 14

1. 0.0246

2. Accept H_1 if > 1 murder in a week.

3. Server utilisation = 0.667
   Av. no. in queue = 2
   Av. busy period = 6 msecs.
   Av. waiting time = 4 msecs.

4. 0.7556 minutes.

Chapter 15

1. $(2\frac{1}{2}, 2\frac{1}{2})$ Z = 30

2. Z = 30,000 (Multiple solns.)

3. Min. blackcurrant = 1
   Nax. elderberry = 4

5. (30,20) Z = 540

Chapter 16

1. Z = 71

2. Z = 218 (2 solns.)

3. Z = 587

4. Z = 430

Chapter 17

1. Max. = 33, Min. = 16

2. Z = 191

3. Z = 67

Chapter 18

1. Value 5/4 geese.

2. a) A(1,0,0) B(0,1,0) V = 0
   b) A(2/3,1/3,0)
      B(0,1/6,5/6) V = 5/3
   c) A(3/5,2/5,0,0)
      B(3/5,0,0,2/5,0) V = 6/5

3. V = 2/9

4. 11/20

Index